DESIGN EXPLAINED

Functional Requirements to Proven Products

By

William Beazley, PhD, PE

Copyright © 2019 William G. Beazley

All rights reserved. Except as permitted by law, no part of this book may be reproduced, stored in a retrieval system, communicated or transmitted in any form or by any means without prior written permission.

All inquiries should be made to the author, Houston, TX USA.

ISBN- 9781795855402

Disclaimer

The material in this publication is of the nature of general comment only and does not represent professional advice. It is not intended to provide specific guidance for particular circumstances, and it should not be relied on as the basis for any decision to take action or not take action on any matter which it covers. Readers should obtain professional advice where appropriate, before making any such decision. To the maximum extent permitted by law, the author and publisher disclaim all responsibility and liability to any person, arising directly or indirectly from any person taking or not taking action based on the information in this publication.

Table of Contents

Table of Contents iii
Table of Figures vii
List of Tables ix
Preface x
Acknowledgements xi
Introduction 1
   Is Mom a Designer? 1
      Go to a restaurant and pick from the menu 1
      Plan a meal from what's on hand 2
      Cook a planned meal 3
      Mom the Designer 3
      Asking for Dinner 3
   Asking for things 4
Requiring What You Want or Need 5
   A Simple Example 6
      What is Missing in the Boss's Description? 6
      Defined Parameter(s) 6
      Agreeing on Measurement Scales 6
      Undefined Parameter(s) 7
      Precision Criteria 7
      Asking the right questions 8
   Prescriptive Product Definition 8
      Defining More Measurable Characteristics 9
      Confirmation by Inspection 11
      The Box as a Prescriptive Product Definition 11
   Work Package Requirement 11
      Resources 13
      Controls 13
      Outputs 14
      The Box Required as a Work Package 15
   Functional Requirements 15
      Defining the Need in a Large Context 16

- Systems Approach 17
- Acceptance Testing 17
- Designing the Solution and its Implementation 18
- Chapter Take-Aways 19
- Problems 19

Specifications 23
- Definition of a specification clause 23
- Metric scales 24
  - Nominal 24
  - Ordinal 25
  - Interval 26
  - Ratio 26
  - Example: Specifying a TV 27
- Criteria 28
  - Allowable Mathematical Operators 28
  - Measurement Scale Comparisons 29
- Specifications Defined 30
  - Specification Clauses 30
  - Venn Diagrams 31
- Process/Procedure Definitions 31
- Specifying Processes and Procedures 32
- Requirements Engineering 32
- Chapter Take-Aways 33
- Problems 33

The Design Process 35
- Design Process Overview 35
  - Proving Performance Using Technology 36
  - Resources 36
- Design Processes 37
  - Analysis and Inversion 37
  - Dimensional Analysis, Scale Modeling and Similitude 38
  - Systems Modeling 39
  - Iteration 40

- Refinement and Decomposition 41
- Standardization of Components 42
- Phasing 44
- Baselining 47
- Chapter Take-Aways 47
- Problems 48

Defining Things That Work 49
- Introduction 49
- Defining Products 49
  - Choice of Representation 49
  - Essential Parameters 50
  - Data of Record 51
  - Rendering of Views and Reports 52
  - Anticipating Implementation in Design 53
- Proving Function 53
  - Analysis 53
  - Assurance 55
- Defining Implementation 57
  - Essential Processes/Procedures 57
  - Assurance 58
  - As-Built Updates 58
- Reporting Designs 58
  - Parts of the Final Report 58
  - Optional Parts of the Final Report 59
- Chapter Take-Aways 60
- Problems 60

Design Disciplines & Technologies 61
- Overview 61
- Engineering disciplines 62
  - Civil Engineers 62
  - Mechanical Engineers 63
  - Electrical Engineers 65
  - Chemical Engineers 66

- Marine Engineers 67
  - Summary of Engineering 68
- Design Technician disciplines 68
- Vendors/Suppliers 69
- Factory/Shop/Construction disciplines and trades 69
  - Manufacturing Occupations 69
  - Fabrication Shop Trades 70
  - Light Construction Trades 70
  - Heavy Construction Trades 71
  - Outsourcing Implementation 72
- Operations & Maintenance Disciplines 72
- Research, Development and Testing Disciplines 73
- Chapter Take-Aways 73
- Problems 73
- Conclusions 75
- Sources Used 76
- Glossary 77
- Index 82
- Appendix A – Writing a Design Problem Statement 83
- Appendix B – Writing a Technical Proposal 84
- Appendix C – Writing a Final Technical Report 85
- About the Author 86

Table of Figures

Figure 1 You are not designing the meal when you pick from the menu. 2
Figure 4 What's in the pantry 2
Figure 4 Available pots and pans 2
Figure 4 Available utensils 2
Figure 5 What's in the refrigerator 2
Figure 8 Collect ingredients and utensils 3
Figure 8 More utensils. 3
Figure 8 Dinner! 3
Figure 9 Remember Mom! (Photo: USDA) 4
Figure 10 I want a box! 6
Figure 11 Possible box dimensions 6
Figure 12 Kilogram No. 20, in the U.S., is one of several "working standards." (Source: NIST) 7
Figure 13 Nominal and dressed lumber sizes. (Source: Voluntary Product Standard PS 20-15, NIST) 10
Figure 14 Benjamin-Moore "Classic Gray" (OC-23) 10
Figure 15 Executing a Work Package will produce a box 11
Figure 16 Elements of a Work Package 12
Figure 17 Dimensioned drawing of part 14
Figure 18 The Boss's Functional Requirement 15
Figure 19 Design Context from Market Need to Work Package 16
Figure 20 Design turns Functional Requirements into proven Product Definitions then into Work Packages. 18
Figure 21 Serving dinner (U.S. Air Force photo/Valerie Mullett) 19
Figure 22 Bicycle parts (Diagram: Wikimedia.org) 20
Figure 23 Doctor discusses a treatment plan with a patient. (U.S. Navy photo by Jason Bortz) 21
Figure 24 Food in the pantry (Photo: Derek Jensen Wikimedia Commons) 21
Figure 25 Confucius (Source: Wikimedia Commons) 23
Figure 26 Nominal Scales have only categories (Image: Wikimedia Commons). 24
Figure 27 Ordinal Scales have order but no uniform interval (Image: Wikimedia Commons), ... 25
Figure 28 A hieroglyphic calendar at Elephantine. (Photo: Théodule Devéria – Wikimedia Commons) 26
Figure 29. Ruler show standard inches and centimeters. (Image: Wikipedia Commons) 26
Figure 30 Typical Selections used to Narrow Product Searches. (Source: Bestbuy.com) 27
Figure 31 Criterion check for carry-on bags. 28
Figure 32 GO/NOGO diameter gauge. (Source: Kaboldy Wikimedia Commons) 29
Figure 33 Universe (U) of all things with the set of acceptable things included. 31
Figure 34 Venn diagram of diesel-powered, personal transportation options 31
Figure 35 Typical failure assessment diagram for pipeline weld flaws 31
Figure 36 If the function that predicts performance can be inverted, then we can calculate the system characteristics needed from the requirements. This can rarely be done. 37
Figure 37 12-percent scale model of the US Navy variant of the F-35 (Source: US Air Force) 38
Figure 38 CSIRO Computer model image of a rogue wave smashing into a semisubmersible platform. (Source: Wikimedia Commons) 39

Figure 39 Iteration Process..40
Figure 40 Decomposition of a bicycle design WBS along its constituent parts. Note that the design has not become refined to more detail. (Source: Garry L. Booker, Wikipedia). 42
Figure 41 Designers prefer COTS products when their rated function meets or exceeds the required function. 44
Figure 42 Systems Development Life Cycle. (Source: US Department of Justice, Wikimedia Commons) 45
Figure 43 Successive baselines converge to a fully acceptable design...............................47
Figure 44 Drafter working in engineering department, 1959 (Source: Seattle Municipal Archives) ...50
Figure 45 Rendering of 3D Shaded View of a Carburetor. (Source: Pixabay)......................52
Figure 46 User interacts with Design using Virtual Reality. (Source: US Air Force)...........52
Figure 47 Prakash 1.5 HP centrifugal pump. (Source: Wikimedia Commons Prakash Worldwide Co. Inc) 55
Figure 48 Designers are engineers, technicians and other professionals who apply principles to prove functionality. 61
Figure 49 Civil Engineers Inspecting a Dam (Source: USACE)...62
Figure 50 Cassie, a walking robot built at Oregon State University Dept of Mechanical Engineering. 63
Figure 51 Wynton Habersham, chief electrical officer for MTA New York City Transit's subway system, showing control panal. (Photo: Metropolitan Transportation Authority / Patrick Cashin.) .65
Figure 52 A Chemical engineer stands in front of the boiler inside the Explosive Destruction System (EDS) Boiler Chiller Container at the Pueblo Chemical Agent-Destruction Pilot Plant. The boiler will provide steam during destruction operations.(Source: PEO ACWA).........................66
Figure 53 Crew members aboard the Auxiliary General Oceanographic Research (AGOR) vessel R/V Sally Ride retrieve scientific moorings located in La Jolla canyon. (U.S. Navy photo by John F. Williams/Released) 67
Figure 54 Sandblasting before painting (source: US Air Force) 70
Figure 55 Worker uses cutting torch. (Source: US Air Force) 70
Figure 56 Stardust Industries light construction worker (USDA Photo by Lance Cheung)...71
Figure 57 Setback levee construction in West Sacramento, CA. (U.S. Army Corps of Engineers photo by Michael J. Nevins)..71
Figure 58 Operations and Maintenance (O&M) personnel often advise designers on best practices. (Photo: US ACWA)..72
Figure 59 Research develops technologies that are applied in design. (Photo: US ACWA) ...73

List of Tables

Table 1 Box Product Definition Requirements 11
Table 2 Box Bill of Materials 12
Table 3 Box Fabrication Process 12
Table 4 Box Work Resource Requirements 13
Table 5 Box Process Controls 13
Table 6 Box Cut List 14
Table 7 Box Process Outputs 14
Table 8 Boss's Functional Requirements for the Original "Box" 15
Table 9 Comparison of Requirement Types 18
Table 10 Comparison of Measurement Scales (Source: Wikipedia) 29
Table 11 Sample Specification Clauses from Each Scale 30
Table 12. Sunshine Cake ingredients 32
Table 13 Analysis of Design Weights to Determine Satisfaction of Weight Specification 54
Table 14 Comparison of STEM Areas by Contribution to Design 61
Table 15 Comparison of Selected Engineering Disciplines 68

Preface

I've been teaching design since the 1970's to students, designers and engineers. Design comes in many forms and disciplines. We all practice it in our daily lives. I've learned that design can be taught and explained in simple terms.

Much confusion results when practitioners and teachers explain design while also explaining their technical disciplines. The main attempts come from the Science, Technology, Engineering and Math (STEM) community. Their Freshman level college text seem unable to explain design independently of their disciplines. But design is a behavior common to many disciplines and exclusive to none.

So, I challenged myself to define design independent of STEM disciplines but showing design applied to their own needs. As a result, we have the reader designing after the first chapter. You are designing now, to get the products and services you need.

We start simple but give you more formal tools you'll need in your career. After the first chapter, you'll strengthen and deepen your design knowledge and techniques. If you are in a project class, you can begin your project after the first chapter, while reading about doing things better. This book gives an integrated view of the multiple disciplines which are applied to design tasks, projects and programs.

This book is an excellent companion to a text introducing your chosen field of engineering, science or technology. There, the designs are more challenging and the designers need to know more. Once you understand the basics, it's easy to acquire the knowledge and experience to specialize.

Finally, this book places many concepts in context to guide future work. It is not intended as a broad survey of applicable research, so I expect interested readers to follow up where their curiosity takes them. I've deliberately made extensive use of Wikipedia in citing sources, hoping you will continue your research there. If I've omitted some important contributions (and there are many), it is not intentional.

Acknowledgements

Thousands have practiced design over the centuries and passed their ideas to colleagues, subordinates and apprentices. With every generation, the explanations have become clearer and more concise. I was fortunate by way of background and circumstances to advance this a step further.

I graduated with a dual major in Mechanical Engineering and Psychology from Tulsa University. I was greatly impressed by ideas on learning theory, metrics, and behavioral objectives outlined by Prof. Irene Horton and many others. As a heavy thermo-fluids and aerospace faculty, they emphasized similitude and modeling in all courses.

Later as a Teaching Assistant, I was allowed to teach Mechanical Engineering Senior Design at the University of Texas by Prof John Allen and Department Chairman J. Parker Lamb. In this course, I introduce assessments of design behavior independent of the required Machine Design content. I continued improving the criterion-based curriculum and assessments when I taught Freshman Design at the University of Illinois.

Over later years of engineering practice in industry I began applying computer tools to support design and drafting, and other process improvements to design work. The work of Richard J. Mayer, President of Knowledge Based Systems, Inc., on applying IDEF family of modeling languages to helicopter overhaul was a big influence my ideas for documenting on-call services such as design.

I began developing training and testing for Professional Piping Designers as Executive Director of the Society of Piping Engineers and Designers (SPED). Pipers design using a host of non-scientific principles and technologies alongside their Piping Engineering counterparts. My many piping designer associates and I described successive levels of piping designer skills from pipe route to senior lead. I trained thousands of pipers and each one challenged me to clarify my ideas for them. It crystalized the distinction of design as a behavior from the principles and technologies used.

I've had a lot of help and encouragement from family and friends. My wife, Merrilee proofread the first draft and several additions. To try to reach a younger audience, I had many helpful comments from my daughter, Andrea Macejak (a middle school teacher) and my granddaughter, Anna Macejak (at 16, an author c). I was particularly encouraged by a cousin, Mary Ellen Johnston Bock, Professor Emeritus at Purdue University, and another professional author and friend, Ian Sutton.

There are so many others that have helped me that I know I've omitted a few key individuals. I am grateful to them as well.

Introduction

There are many books about design. Few answer the one question most people ask, "What is design and how do I do it?" Design defines proven products and services to function as needed. The rest of this book explains this in more detail.

We will provide the details quickly. After the first chapter, you should be able explain design and give a few examples from your own experience. In fact, after the first chapter, you could start a small, classroom-style design project for a semester length class. You'll see that you've been using design principles already and so have others around you. We just make it a little more formal.

In the remaining chapters, we will help you be a better designer. Chapter 2 will go deeper into specifications, the language of requirements. Chapter 3 will explore the design process itself and the methods designers use to get solutions to problems. Chapter 4 will re-state design process using the new, more technical terms we've introduced. Chapter 5 will give overviews of specialized fields of design and the technical fields that support them.

You design things already to satisfy your own needs. We will give that a little more structure that you'll find helpful. Let's give a brief example from your own home.

Is Mom a Designer?

It's getting close to dinner time and Mom is in charge. Mom consider her options:

1. Go to a restaurant and pick from the menu
2. Plan a meal from what's on hand
3. Cook a planned meal.

Let's look at each option and what might result

Go to a restaurant and pick from the menu

The family goes to a restaurant and picks from what they offer from the menu. Every choice is described in enough detail that a typical diner knows what they are selecting. So, all they have to do it "name" the meal they select, i.e., "I'll have the roast beef diner." They hope they like what they get and mostly they do. There are a few things they didn't like or couldn't eat.

What is Design?

Design defines proven products and services to function as needed.

…you've been using design principles already and so have others around you…

Was mom a designer here? Hardly. After satisfying their general need for dinner, they just chose existing dishes by their name and

Figure 1 You are not designing the meal when you pick from the menu.

description in the menu. Most of the choices turned out well but a few didn't.

Plan a meal from what's on hand

Mom plans out a dishes based on what is on-hand in the pantry and refrigerator. She thinks about what each member of the family likes to eat and their allergies. She also plans how to cook the dishes in the kitchen with the utensils she has. In her mind, she works though all the steps needed to cook the dishes and puts them in the "meal plan". With enough time, she can plan all the meals for the next week.

Figure 4 What's in the pantry

Figure 4 Available pots and pans

Figure 4 Available utensils

Figure 5 What's in the refrigerator

Is Mom a designer here? Yes. Mom has considered the needs for dishes and planned a meal to meet them. By carefully considering everyone's needs there is a good chance that, when it is cooked, it will be a satisfying meal.

Cook a planned meal.

Mom uses the meal plan to gather all the ingredients she needs for the dishes from the pantry and refrigerator and all the utensils to cook with. Then she cooks the meal.

Figure 8 Collect ingredients and utensils

Figure 8 More utensils.

Is Mom a designer here? No. She's a cook making dishes according to a plan. If the meal is well planned and everything needed is on hand, everyone should be satisfied. If Mom has planned several meals, she can cook them on different nights and use them over and over.

Mom the Designer

We see Mom in several roles:

- Mom as a restaurant diner, picking from a menu of predefined meals,
- Mom as a designer of dishes and meal plans to cook them,
- Mom as a cook carrying out a meal plan

Figure 8 Dinner!

We recognize that something special took place when Mom designed the dishes and planned the meal. She thought about what needs of the family and the constraints of the ingredients and utensils available. She designed dishes and a meal plan she knew would meet those needs using available ingredients and utensils.

Asking for Dinner

Now, let's see how people in the family ask for their dinner. They might say:

- **Describe it**: "Mom, let's order a sausage and pepperoni pizza tonight."
- **Make it from an existing plan**: "Mom, could you cook up your special meat loaf again tonight?"
- **Design it as needed, then make it**: "Mom, could you think up something for tonight that we'd all like to eat and tell us how to cook it?"

Each way the meal is asked for gets a different result. The first one is a description of what to get. The second is a request to cook (make) it according to an existing plan. The third is a statement of need with a request to design the meal and, then, explain how to cook the meal.

Asking for things

When people need or want something, they ask for it. If it exists, they ask for it by name or by its characteristics. If nothing described appears to meet their needs, they will have it "designed" to their requirements and then built to that design.

People who "design" to meet requirements are referred to as "Designers." Design is common in our everyday lives. Many people meeting our wants and needs are doing "design" whether it's called that or not.

The rest of the book just makes these ideas more formal. We'll begin using terms common in Science, Technology, Engineering and Mathematics (STEM). Design is not constrained to these disciplines but by the act of defining proven products or services to meet functional needs. If you get lost, just think of Mom…

3 Ways to Ask
1. **Describe it precisely,**
2. **Tell how to build it,**
3. **Ask a need be fulfilled.**

Figure 9 Remember Mom! (Photo: USDA)

Requiring What You Want or Need

To formally specify as compulsory is to "require". We will find that there are ways of requiring things that lead to different results with each with its own sets of advantages and disadvantages. Designer are only interested when we require our needs be satisfied. Designers will define a product or service that meets those needs.

> "That, if requiring fail, he will compel"
> Duke of Exeter threatening the King of France to give up the crown in Shakespeare's <u>History of Henry V</u>

The three general ways something can be specifically "required" are by:

1. **Prescriptive Product Definition**, as for an item defined by its measurable characteristics;
2. **Work Package**, ask that a process be done that creates the item, and;
3. **Functional Requirements**, ask for something that performs certain functions.

If you ask for something in any of these three ways, you will usually get it.

Product Definition	Work Package	Functional Requirements
1. 1/4 in. - 20 TPI x 1 in. Zinc-Plated Coarse Thread Carriage Bolt 2. 1/4 in.-20 TPI Zinc-Plated Hex Nut	Grandma's Cookie Recipe	The US "should commit itself to achieving the goal, before this decade is out, of landing a man on the Moon and returning him safely to the Earth." John F. Kennedy (May 25, 1961)

By using the term "prescriptive," we reinforce the idea that the defined characteristics MUST be present in the required product or service. If you ask for something by its product definition, then a vendor or supplier will merely checks off each characteristic in fulfilling your request. If no one has it, then your request goes unfilled.

Sometimes you have a need to fulfill and the product is not known. Then, you can ask for that product to be designed to meet your specific need. More formally, **Designers** utilize **Functional Requirements** to create proven **Product Definitions** and the **Work Packages** that implement them.

> **Designers** utilize **Functional Requirements** to create proven **Product Definitions** and the **Work Packages** that Implement them.

When composing these requirements, a simple request generally moves toward on of these forms. We'll begin with a simplified example of how these forms take shape and what concepts are used to make them clear..

Figure 10 I want a box!

What's Missing?

- Where Dimensions Apply
- Measurement Scale/Units
- Wall Thickness
- Shade of Gray
- Material of Construction
- Closed or Open
- Tolerance on Dimensions

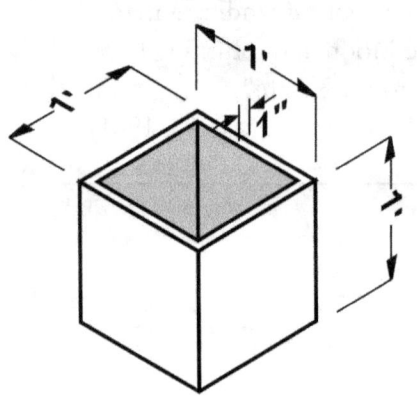

Figure 11 Possible box dimensions

A Simple Example

Your boss walks into your office and declares (Figure 10),

"I need a gray, 1ft by 1ft by 1ft box."

That sounds simple enough. Well, let's think about what he's asking for.

What is Missing in the Boss's Description?

There are several problems with the boss's description:

- We don't know if the dimensions are to the inside or outside of the box;
- The thickness of the walls is not given;
- The exact shade of gray is not defined;
- Other parameters are not defined; and,
- The tolerance of dimensions and other parameters are not given.

We'll use these issues to highlight the how designers think about problems.

Defined Parameter(s)

Since all the stated measurements are the same, our boss didn't bother to explain which is the **length**, **width** or **depth** of the box. Is he referring to the **inside** dimensions or **outside** dimensions? Does he care how thick the walls are?

Normally, these questions are resolved with a sketch or drawing. Because a box is a physical object, a drawing showing all the relevant parameters is made (see Figure 11). Drawings usually contains one or more "views" so that all features are clear. Solid and dashed lines show features that are visible or hidden in each view. "Dimensions" are applied to each feature to show the parameters of that feature. Callouts and notes determine other relevant features.

Agreeing on Measurement Scales

Additionally, the boss gives the dimensions of the box in "feet." In the US, feet are units of standard length. The **foot** (pl. **feet**; abbreviation: **ft**; symbol: ' <the prime symbol>) is a unit of length in the imperial and US customary systems of measurement. Since 1959, both units have been defined by international agreement as equivalent to 0.3048 meters exactly. In both systems, the foot comprises 12 inches and three feet compose a **yard**.[1]

---

[1] https://en.wikipedia.org/wiki/Foot_(unit)

It's important that his dimensional requirement is in a **standard unit of measure** that every potential supplier can be expected to use[2]. A **standard unit of measure** is an agreed unit that has a physical standard or repeatable defining derivation that everyone can replicate for their own use (see Figure 12). Length dimensions use a type of measurement scale that has a constant unit separating its intervals and a true zero. This type of measurement scale allows us to apply certain criteria to it, such as a tolerance, e.g. $\pm 1/8$".

Figure 12 Kilogram No. 20, in the U.S., is one of several "working standards." (Source: NIST)

Gray, is a little less standard. Gray is an intermediate color between black and white. It is a neutral color or achromatic color, meaning literally that it is a color "without color."[3]

It also matters how the "color" is generated. Four common ways to generate color is by:

1. Reflection (Inks and paints),
2. Emission (fluorescing phosphors or Light Emitting Diodes)
3. Filtered Transmission (clouds)
4. Refraction (clouds).

So, the source of Gray color matters.

Undefined Parameter(s)

There are also features not define.d by our boss. What **material** is the box going to be made of? Foam? Metal? Wood? Plastic? Cardboard? Some materials might meet his purposes, some may not.

Finally, the color is specified but the **surface containing the color** is not. Plastic can be tinted. Wood can be painted, Metal can be powder coated or painted. Cardboard can be printed or dyed. Different materials present different color options Since the box will probably be paintend or the material tinted, it will a reflected gray.

Also, is this box **closed or open**? If closed, **how**? Is it to be glued or nailed shut? Is is closed with a lid? What does he need?

Possible Box Materials
• **Foam?**
• **Metal?**
• **Wood?**
• **Plastic?**
• **Cardboard?**

Precision Criteria

Isn't one foot always one foot? Can it be one inch more and still be acceptable? How about an inch less? How close must we get to one foot and still make the boss happy? This range of allowed results

---

[2] The United States is the only industrialized nation that uses the international foot and the survey foot (a customary unit of length) in preference to the meter in its commercial, engineering, and standards activities. See: https://www.cia.gov/library/publications/the-world-factbook/appendix/appendix-g.html *The World Factbook* 2018. Washington, DC: Central Intelligence Agency, 2018. Retrieved February 4, 2018.

[3] https://en.wikipedia.org/wiki/Grey , accessed December 31, 2018.

establishes the **"tolerance"** on the measurement. **Tolerance** gives us a **criterion** for the precision on a measurement.

The tolerance criteria automatically tells us how to **"inspect"** the delivered box. The supplier won't ship until the product meets the tolerances. The buyer won't approve payment until the delivered product meets tolerances.

Each measured parameter might have its own tolerance or all parameters or the same type might have the same tolerance. (All Dimensions ± 1/8"). Other criteria on measurements might be different, depending on the measurement and scales used. We will explore this more, later.

> **Tolerance** gives us a **criterion** for the **precision** on a **measurement**…

Asking the right questions

It's time to go back to the boss with questions. His answers might fill in the blanks or even change his requirements altogether. He can add requirements, i.e., make the list longer and more complete. He can clarify his requirements by showing their location, application or by adding tolerances.

He can also delete or modify requirements that are inconsistent. For example, powder coating cannot be required for cardboard and Red-Green-Blue (RGB) emitted color schemes are not appropriate for reflecting color pigments.

He can also make a decision that restricts the box to a class of boxes (cardboard) and "refine" the requirements to be appropriate for that design choice. The more "abstract" requirements have now become more "detailed" by this refinement.

This process of interacting with the client to clarify and refine requirements while adding more detail to the solution is called iteration. Requirements and design details can grow until success is achieved (or not). During this process, entire approaches or major design candidates might become unfeasible and be abandoned. Knowledge and experience help the client and designer avoid wasted efforts and dead ends

> **Requirements can be:**
> - Completed
> - Changed
> - Added to
> - Clarified
> - Constrained
> - Sub-Divided
> - Refined

Prescriptive Product Definition

Once we get all the relevant parameters, measurement scales and criteria settled, we can write a **Prescriptive Product Definition** requirements (**Product Definition** for short). A product definition is a set of requirements that define the product characteristics in ways that can be directly measured.

> A **product definition** defines the product by **characteristics that can be directly measured.**

We are using "product" in a rather general way, i.e., any product or service that can be asked for by measurable characteristics. Designers, however, are only interested in products which satisfy needs. If anything can be found that satisfies all these measurable characteristics, then it is, by definition, the "product."

Defining More Measurable Characteristics

Let's add a little more detail to the box to complete its measurable characteristics. This includes the material and the full paint spec.

If we require a material to be "oak", then we can inspect it and confirm it's oak. There can be many species of oak and many colors, but the material is either oak or it's not. We can always go to the product and confirm it.

> **We can always inspect and confirm the product characteristics.**

Length might be another measured characteristic. If a pipe is to be 10 inches +- 1/8 inch, then we can measure that. If it's required to be "at least" 10 inches, we can measure that also.

Some standardized sets of characteristics are referred to by **names**[4]. For example, lumber comes in several **"nominal sizes"** (see Figure 13). Nominal lumber sizes are often "dressed" to a lesser size to remove splinters and sharp edges, and to allow for shrinkage during drying. Tolerances on these dimensions are rarely significant in practical applications and are generally ignored. (please note that "nominal" sizes here are more like the **Ordinal Scales** we discuss later, since as nominal sizes increase so do their dressed sizes, in order.)

> **Some characteristics are standardized and referred to by their names.**

---

[4]American Softwood Lumber Standard, http://www.alsc.org/greenbook%20collection/ps20.pdf, accessed 3 December, 2018.

Item	Thicknesses					Widths				
	Nominal Inch	Minimum Dressed				Nominal Inch	Minimum Dressed			
		Dry a		Green a			Dry a		Green a	
		mm	inch	mm	inch		mm	inch	mm	inch
Boards	3/8	8	5/16	9	11/32	2	38	1-1/2	40	1-9/16
	1/2	11	7/16	12	15/32	3	64	2-1/2	65	2-9/16
	5/8	14	9/16	15	19/32	4	89	3-1/2	90	3-9/16
	3/4	16	5/8	17	11/16	5	114	4-1/2	117	4-5/8
	1	19	3/4	20	25/32	6	140	5-1/2	143	5-5/8
	1-1/4	25	1	26	1-1/32	7	165	6-1/2	168	6-5/8
	1-1/2	32	1-1/4	33	1-9/32	8	184	7-1/4	190	7-1/2
						9	210	8-1/4	216	8-1/2
						10	235	9-1/4	241	9-1/2
						11	260	10-1/4	267	10-1/2
						12	286	11-1/4	292	11-1/2
						14	337	13-1/4	343	13-1/2
						16	387	15-1/4	394	15-1/2

Figure 13 Nominal and dressed lumber sizes. (Source: Voluntary Product Standard PS 20-15, NIST)

So, if we specify a 1 in nominal lumber thickness, we can expect to receive lumber dressed to a minimum ¾ in thickness. That means the ¾ in thickness might vary +1/4-0 inch.

Defining a Coat of Paint:
- Base Paint Used
- Color Tint
- Thickness of Coating

Coatings are generally applied to material if they are not already tinted. Coatings are applied as paint, powder-coated, plated, etc. Paint is usually applied after a primer and the resulting overall thickness. In the U.S. the thickness of paint is expressed in mils (one mil equals 1/1000 of inch). The rest of the world expresses coating thickness in microns (1 micron = 1 millionth of a meter and 25.4 microns = .001" inch or 1 mil).

Here is a typical paint spec from Benjamin-Moore[5]:

1) 1st Coat: Benjamin Moore Fresh Start Multi-Purpose Primer N023 (44 g/L), MPI # 6, 17, X-Green 17, 39, 137, X-Green 137, LEED Credit, CHPS Certified.
2) 2nd Coat: Coronado Rust Scat Waterborne Acrylic Gloss 80 (224 g/L), MPI # 114, 154, LEED Credit.
3) 3rd Coat: Coronado Rust Scat Waterborne Acrylic Gloss 80 (224 g/L), MPI # 114, 154, LEED Credit.

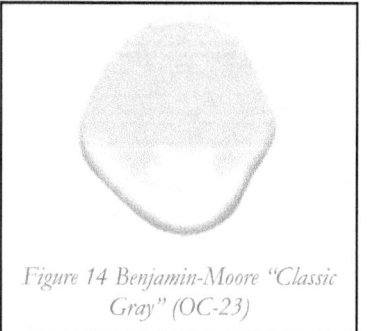

Figure 14 Benjamin-Moore "Classic Gray" (OC-23)

As for colors, paint manufacturers will all supply pallets of colors to choose from. From the same manufacturer, we will pick "Classic Gray" (OC-23). Benjamin-Moore's "Classic Gray" is shown in Figure 14.

---

[5] https://www.benjaminmoore.com/-/media/sites/benjaminmoore/files/pdf/commercial-us-6_6.docx

Confirmation by Inspection

The most important benefit of prescriptive product definition is that they can be **confirmed by inspection** or observation without the design functioning. Inspection measures the characteristic on a product and compares it to the criterion specified. By confirming the required measurable characteristics and methods of manufacture, the quality of the design itself is assured.

> **Product Definition**
> **Most important benefit**: Can be confirmed by inspection or observation without the design operating.
> **Biggest weakness**: Suitability for any particular purpose is not guaranteed.

The weakness of prescriptive requirements is that the suitability of the design for any particular purpose is not guaranteed. That assurance is the job of the designer. He must consider the functional requirements of the design and convert that to prescriptive requirements with a justification.

The Box as a Prescriptive Product Definition

Let's restate the boss's request in terms of more measurable parameters. (see Table 1). The level of detail has increased over the boss's initial request. In other words, we have "refined" the initial product definition to the one shown in Table 1. With the boss's agreement, anything meeting these measurable characteristics is, by definition, "what he wants."

This product definition is handy if we seek to buy the product off-the-shelf. But there is another way to get a specific box: Have it made.

Table 1 Box Product Definition Requirements

Type: Box
  Closed on all six sides
  Separate Closable top
Delivered Configuration: Unclosed
Dimensions:
Inside:
  Width: 1ft ±1/8in
  Height: 1ft ±1/8in
  Depth: 1ft ±1/8in
Wall Thickness: 1in nominal
Material: Oak
Coating:
Primer (1 coat) Benjamin Moore Fresh Start Multi-Purpose Primer N023
Paint (2 coats) : Coronado Rust Scat Waterborne Acrylic Gloss 80 (tinted "Classic Gray" #OC-23)
Coating Thickness: 3 mils or greater

Work Package Requirement

Another way for the boss to get what he wants is to completely explain how to create it (see Figure 15). We call these complete set of instructions a **Work Package.** A work package, if executed, will create our box.

Generally, a work package has 5 elements:

1. Inputs Required
2. Process(es)
3. Resources Required
4. Controls
5. Output

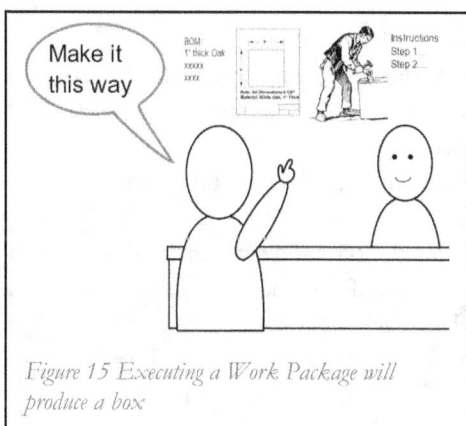

Figure 15 Executing a Work Package will produce a box

These elements are illustrated in Figure 16.

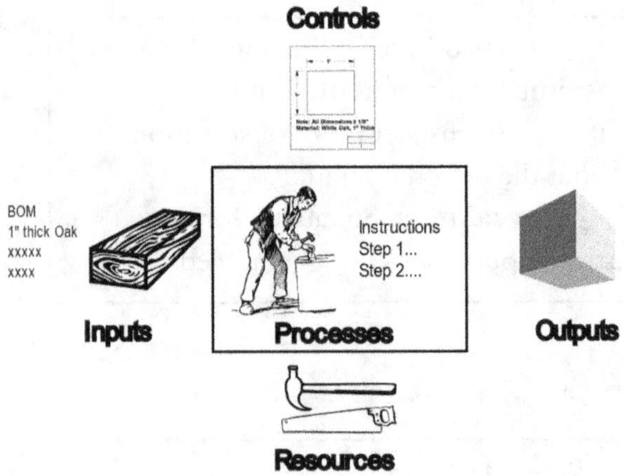

When a work package is required, all the elements are required as well.

Figure 16 Elements of a Work Package

Input Requirements

Inputs represent the items that are "processed" into the output. In the case of our box, we will take wood, nails, glue and paint and output a box. Each time the work package is executed, a fresh set of inputs is required, and a new output is generated.

Importantly, certain sizes, quantities and characteristics of input items are required. If the inputs are physical, the inputs are sometimes referred to as a **Bill of Materials** (**BOM**).

The BOM for our box is shown in Table 2). All of the materials needed to build the box are in the BOM. In addition, some items (brushes and drop cloth) are consumed but are not "part" of the final output. Items consumed in the process are generally shown in the BOM because they are procured and gathered together with all other materials needed by the process.

Processes of a Work Package

Work Package Processes are another type of prescriptive requirement but it's on the process of creating the "**end product**". (Here, the end product is the output of a work package designed to create it.) It can be on the steps to be taken, the conditions of manufacture, when and how the results are to be confirmed. It must be observed directly by the client or certified by a trusted manufacturer or fabricator. Process steps for fabricating our box is shown in Table 3.

One common example is heat treatment. When two pieces of metal are welded together, the alloy and crystalline structure of both pieces are altered by the heat of welding. The desired properties of the metallic pieces is restored by heat treating. The design will call out a standard heat treatment, which dictates the temperature then metals are to be raised to, how long to hold that temperature and how slowly to cool or quench the pieces. Either the client will observe the heat treatment process directly or he will accept a document from the heat treater certifying that the specified process was carried out.

Frequently, a manufacturing process will include "**hold points**." All subsequent steps in the process are put on "hold" to await an

Table 2 Box Bill of Materials

1. 7ft. Oak Board, 1 in thick, 13-½ in wide.
2. 8 oz Wood Glue
3. 36 #15 x 1-1/2 in. 4-Penny Bright Steel Finish Nails
4. 2 oz Wood Filler
5. 1 Quart Benjamin Moore Fresh Start Multi-Purpose Primer N023
6. 1 Quart Coronado Rust Scat Waterborne Acrylic Gloss 80 (tinted "Classic Gray" #OC-23)
7. 1 each Drop Cloth
8. 2 each Disposable Paint Brush

Table 3 Box Fabrication Process

1. Confirm fulfillment of the BOM
2. Cut the sides, top and bottom of box
3. Dry fit all pieces, adjust or replace as needed.
4. Glue and nail 4 sides and bottom, while clamped and square.
5. Fill nail holes and set aside to dry.
6. Sand and Prime box and set aside to dry
7. Paint 1st coat and set aside to dry
8. Paint 2nd coat and set aside to dry
9. Inspect box as meeting final requirements for output

inspector to confirm the quality of the work. Building inspectors frequently require hold points so that plumbing or electrical work can be inspected before it is covered up by additional construction. These process requirements are part of most building codes.

Resources

Other **Resources** are required for a Work Package process to execute. These include:

- Tool and machinery time
- Labor skills and work time
- Calendar Time (duration)
- Assembly/Staging area time

Table 4 Box Work Resource Requirements

General Carpenter (5 Hours)
Carpenter Tools (as needed)
Square
Measuring Tape
Hammer
Putty Knife
Table Saw (1 Hour)
Work table (1 Hour x 5 sessions)
Shelf Space for drying (4 Sq. ft. x 24 Hours)

Resources are "assigned" in the sense that while they are used, they are not longer available to other processes.

A set of work resources that might be required by our box creation are shown in Table 4. Time on equipment, work and shelf space are all needed to make the box. Note also, the labor of a skill carpenter is also needed as are carpenter tools. Generally, trades supply their own tools but a table saw might belong to the shop. While all these things are being used to build the box, they can't be used on any other job.

Controls

Controls are tools used to assure proper execution of the work package processes. Common control are:

- BOM checklist, used to acquire and confirm all the input materials are on hand.
- Engineering drawings showing how to manufacture and assembly the output of the process.
- Required inspections within and at the end of the process

Table 5 Box Process Controls

BOM Checklist
Parts Cut List
Assembly Drawings / Instructions
Final Inspection Checklist

Process controls for our box are shown in Table 5.

One of the sub-tasks in our box is the cutting of six sides of the box from the 7 ft board. We can choose to have a scaled drawing detailing each part (see Figure 17) or a "cut list" with the relevant dimensions to be cut (see Table 6).

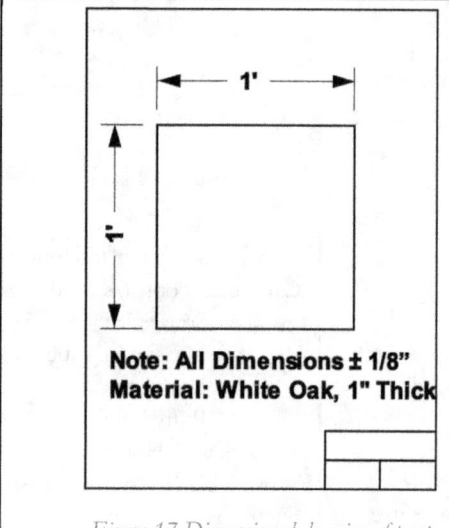

Figure 17 Dimensioned drawing of part

Table 6 Box Cut List

Item	Width	Length	Thick
1	12-3/4	12-3/4	3/4
2	12-3/4	12-3/4	3/4
3	12-3/4	12-3/4	3/4
4	12-3/4	12-3/4	3/4
5	12	12	3/4
6	13-1/2	13-1/2	3/4

The final result would be a box with the same characteristics of the measured requirements. Thus, a checklist would be a control on the final verification step of the process.

Outputs

Each process has defined results, the **Outputs**. There could be many outputs, some desirable and some not. For example, one process might produce a part and also produce scrap (waste). The part might move on to other processes, while the scrap might move to recycle or disposal.

In this overarching process, one output should have the same characteristics as our measured requirements stated previously plus additional outputs of scrap wood, sawdust, dirty brushes and tarp for disposal.

We might say that the process is designed to produce boxes "rated" as described by the measurable requirements. Give it the inputs and resources, start it up and you get the box (plus the scrap). The process outputs for our box is shown in Table 7.

Table 7 Box Process Outputs

1. Box Meeting Stated Requirements
2. Scrap
3. Sawdust
4. Dirty Brushes

The Box Required as a Work Package

If all the elements of the box work package are specified and satisfied, the output should result, i.e., the box. We won't restate all the elements here but for most products, the work packages can be quite detailed. Depending on their scope and area of application, they are known by many names:

- Recipes
- Treatment plans
- Construction Contracts
- Procedures
- Work Proposal

They are part of our everyday lives.

Functional Requirements

So now you ask your boss, "what do you need this for?" "Oh," he says, "we need to ship our new stuffed bears to our stores." (See Figure 18.) The "box" has a function. It's the function that fills his need.

We call this need, the **Functional Requirements** for the object requested. It determines what the object must do, i.e., how it must perform its function. Anything that can be shown to function as required is an acceptable solution to the boss's need.

Figure 18 The Boss's Functional Requirement

So, the boss thinks for a minute and outlines the functional requirements for the box in Table 8. These requirements never mention the word "Box." In fact, the final form of packaging might not be "cubic" at all. Many forms of packaging might meet these requirements.

Table 8 Boss's Functional Requirements for the Original "Box".

Functional requirements seek only to dictate that the design will perform at or above certain levels. The system in question must be operating in the state or sequence of states (process) the requirement is to be measured. This can be the normal operating state/process, an extraordinary operating state/process or one of several shutdown states/processes. To prove performance, the object might be measured while running in that state, if it cannot be simulated first and confirmed later.

New Stuffed Bear Product Packaging Functional Requirements:
• Contain a bear approximately 8"x8"x8" uncompressed.
• Product compression for packaging of 1/8 inch on all sides is acceptable. Compression cannot exceed ¼" all sides:
• Allow for 1" of protective packing on all sides.
• Estimated shipping weight of bear is 3lbs.
• Color and label the package consistently with the corporate retail color scheme, which is blue.
• Meet US Postal Service packaging requirements for First-Class Package International Service

In the boss's case, his packaging must work in two states: Shipping and Retail. There is a chance that a single container will work for both states, otherwise, a package within a package will work. It might be two packages: An outside one to take the rough abuse of shipping and a more attractive one for retail display.

Defining the Need in a Large Context

In writing functional requirements, the client must consider carefully what performance is needed from the design to satisfy a market need. The **context for the design** is a series of analyses linking a Market Need through Functional Requirements. The total context is shown in Figure 19.

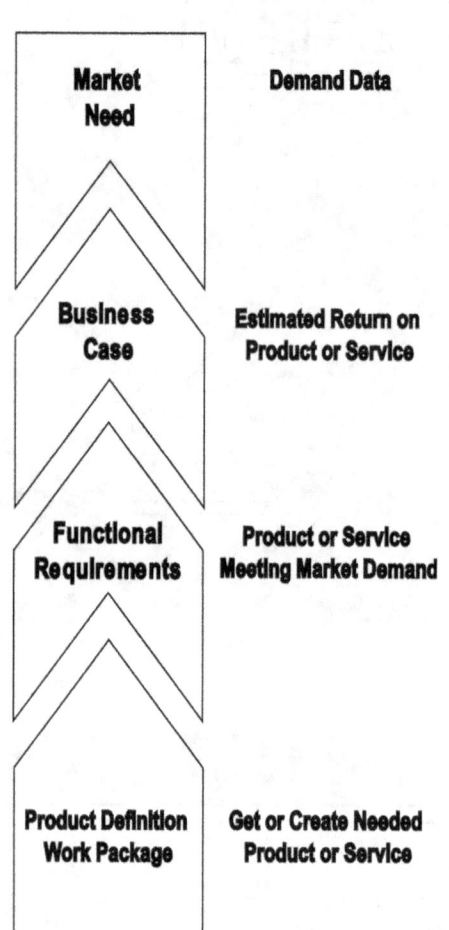

Figure 19 Design Context from Market Need to Work Package

Commercial clients generally begin with a Market Need for a product or service that provides a context for the design in increasing detail:

- **Market Need** is a generalized statement, with supporting data, that a need exists for a product or service strong enough for it to be sold;
- **Business Case** is an estimated return on the product or service that satisfies the market need;
- **Functional Requirements** describe the functions the needed product or service performs; and
- **Product Definition** or **Work Package** define how to get or make the product or service.

Businesses exist to provide products and services that satisfy market needs. Once they have data that identifies a need, they try to make a business case that justifies designing a product or service to satisfy it and provide a return to investors.

The business case analysis shows that a product with the desired performance will improve profits and make a return on the investment. It will include:

- an analysis of market demand for the product,
- a budget for the design and manufacturing effort,
- a schedule, and
- an analysis of expected benefits of the design are realized.

Any part of the design context can and frequently does change as the design proceeds. This usually results in a change to the business case. For big capital projects, the business case is rechecked frequently, particularly before another level of capital must be committed to proceed.

The main point is that the required functional performance is always in a larger context that justifies the work. It is that justification that provides the funds. It includes considerations the designer cannot control and is the ultimate performance (economic and functional) required of the system.

Systems Approach

The main advantage of functional requirements is clear when taking a **Systems Approach to Design**. If you know how the parts work together, you then set the requirements for each part individually. The performance of the larger system is proven using the proven performance of smaller subsystems.

> Systems Approach: If you know how the parts work together, you can set the requirements for each part individually.

For example, a large chemical plant is comprised of hundreds of smaller supplied subsystems, components, and other equipment. Engineering teams have analyzed the overall plant and how each subsystem must perform. When subsystems meet their individual requirements, the system will work as required. When the go-ahead is given, the subsystem equipment is procured, assembled, performance-tested to meet expectations.

In turn, suppliers aggressively design products and systems that outperform their competitors to be chosen for the next system design. Knowing higher performing products exists allows system designers to demand more from their vendors and offer more to their clients. Overall performance increases.

Acceptance Testing

Acceptance Tests are a battery of trials confirming expected functional performance. Acceptance test put the system in a defined set of states or sequences where performance is predicted and confirmed. Acceptance testing can be a single test or several individual tests over days and months.

A special case of this is the **Pilot Test or Prototype Test**. A smaller (pilot) or preliminary (prototype) version of the end system is built and subjected to extensive testing. Such testing frequently uncovers design errors or opportunities to add value to the final end system. Test can exceed desired functionality and can lead to system failure. Tests to destruction can reveal how robust the system is to failure and how operators can react to it safely.

> Pilot and Prototype testing frequently uncovers design errors or opportunities to add value to the final end system.

Designing the Solution and its Implementation

Now, we are ready to answer, "What is design?" Design turns functional requirements into proven product definitions then into work packages to build and implement them. (See Figure 20.) In the special case that there is an off-the-shelf solution available to the designer, then the specified **product definition** and implementation work package is very simple or not stated.

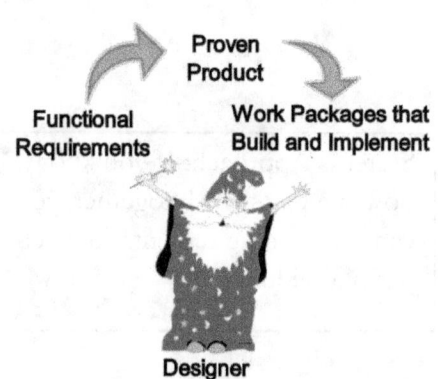

Figure 20 Design turns Functional Requirements into proven Product Definitions then into Work Packages.

Importantly, Functional Requirements define only the need not the solution. **Designers** are specialists skilled at defining solutions and proving functionality. Frequently, they also design the work package that will create or implement the defined solution.

Of course, it is useless to define a solution without some clear reason to be believe it will function as required. That's where the "secret sauce" is for designers: Good designers are good at picking product characteristics and also good at "proving" they function as required. Better designers also define products that are easy to implement, i.e., their work packages are low in cost, short in time frame and fit the resources available.

> **Good designers** are good at picking product characteristics and also good at "proving" they function as required. **Better designers** also define products that are easy to implement.

Frequently, however, the designer defines a product that already exists. If so, it can be bought **"Commercial Off The Shelf"** (**COTS**) and implemented directly or with minor adjustments or accommodations. It's almost always cheaper and faster to buy COTS products than to make it yourself. Designers frequently do a make-or-buy tradeoff study to decide what to do.

All three types of requirements realize our box in different ways (see Table 9). Only Designing to a Functional Requirement has any guarantee of meeting the needs of the customers, provided we don't

Table 9 Comparison of Requirement Types

Type	Realization	Pros	Cons
Product Definition	Procurement or Work Package	Confirmed by inspection COTS Possible	Utility not Guaranteed
Work Package	Execution	COTS not required	Utility not Guaranteed
Functional	Design & Implementation	Guarantees Utility Fewest Limits on Solution Confirmed by Acceptance Tests	May be Impossible or Unfeasible

run out of resources before the design is completed and the solution implemented.

Chapter Take-Aways

We've introduced three ways to ask for a product or service:

- Product Definition
- Work Package
- Functional Requirement

If you ask by product definition, you can confirm it by measurable characteristics but can't guarantee if will fulfill any functional need.

If you know what you need, a clever designer can define a proven product that will satisfy your functional requirements. There is a risk that no technology will be found that will perform as needed.

Designers can also design a work package that will build a product. Work packages are just another designed product, but they require inputs and resources to execute.

Finally, product definitions measure characteristics, while functional requirements measure performance. Both need to be very specific to be useful, because they precisely determine what defines the characteristics or functions of what you get.

Problems

1. Mrs. Jones tells her husband that she is tired of walking around the wall that separates the kitchen from the dining room: She would like a shorter path between the two for serving food. Mr. Jones examines the wall and decides it is needed. Mr. Jones asks his wife to choose between removing the wall or installing a serving bar in the wall. She picks removing the wall. A carpenter prepares a bid to remove the wall and fix up what remains. Mr. and Mrs. Jones accept the bid, the work is completed, the site is cleaned up and the work is paid for.
    a. Who set the functional requirements of this design? Why?
    b. Who designed the "product" definitions of the two possible solutions? When one was selected, did this choice limit the requirements to be met by the carpenter?
    c. Who designed the work package that executed the chosen solution? Why?
    d. Why was the fact that the wall was not load-bearing critical to deciding that a structural engineering analysis was not needed to prove the design safe?

Figure 21 Serving dinner (U.S. Air Force photo/Valerie Mullett)

2. Bob decides he wants a new bicycle. He wants to ride of trails with rough terrain and far from possible repair stations, so wants the highest possible reliability on the trail. He narrows his choices to a mountain-style bike with 2.1" x 26" tires and shock absorbers on the front wheels. Then he reads the user ratings online to eliminate the "bad actors". Finally, he picks the bike with the highest quality components, methods of construction and longest warranty. He purchases the bike and assembles it. On his first ride a brake caliper fails and is replaced by the dealer. He has no further problems during and after the warranty period. He currently uses and enjoys the bike.

   a. How did Bob define his functional requirements for the bike?
   b. How did Bob narrow the possible solutions by making some "design" decisions?
   c. What characteristics did Bob evaluate as a proxy (representative measure) for reliability of design?
   d. How did Bob attempt to "Maximize" expected reliability in his final choice?
   e. What were the steps in Bob's implementation of the design?
   f. Did his implementation involve a kind of "acceptance test?" How was acceptance confirmed by Bob?

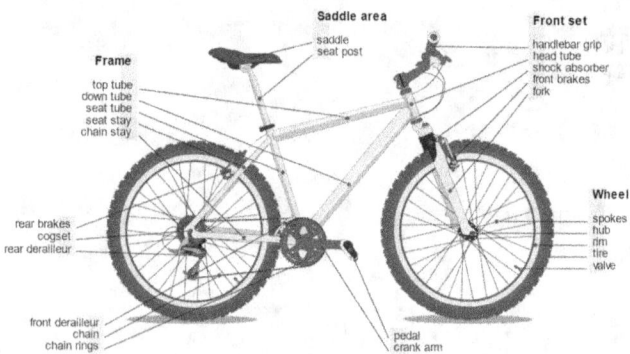

Figure 22 Bicycle parts (Diagram: Wikimedia.org)

3. You go to the doctor because you are sick and need to feel well again. The doctor asks you about your symptoms, examines you and runs some tests. He announces his diagnosis as to the probable cause of the symptoms. Then he creates a "treatment plan" that he expects will relieve your symptoms and cure your illness. He writes a prescription and explains how to take the medicine and other steps that will help. He asks you to return in 7 days for a follow-up.

   a. Who has stated the initial need?
   b. Is the diagnosis a refinement of the initial need or a solution that satisfies it?
   c. Is the treatment plan a solution to the need? Explain why?

    d. Is the treatment plan comparable to a "work package"? Why? If so, what are the inputs, outputs, processes and controls
    e. After the treatment plan is followed by the patient (you), why would the doctor ask you to return for a follow up? What role is he playing in his own "work package?"

4. Mom has to fix dinner for the family (5 people). Dinners is in one hour. Ingredients have to come from the refrigerator or pantry. Cooking in the kitchen using utensils on hand. The father hates broccoli, one daughter won't eat ham and one son won't eat potatoes.
    a. What are the functional requirements?
    b. What resources, time, options, tools have been named? Which options have been excluded?
    c. If she decides what to fix, is it fair to call the meal a "defined product"? Why?
    d. If she decides how to cook the meal in full detail, is it fair to call it a "Work Package"? Why?

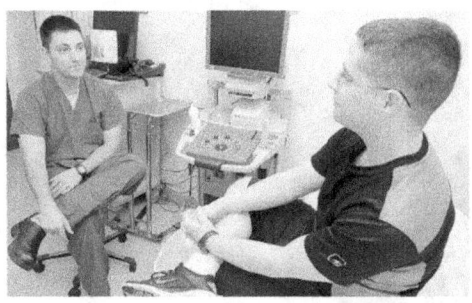

Figure 23 Doctor discusses a treatment plan with a patient. (U.S. Navy photo by Jason Bortz)

Figure 24 Food in the pantry (Photo: Derek Jensen Wikimedia Commons)

Specifications

We have introduced requirements for both characteristics or functions of a product or service. **Specifications** (**Specification Clauses**) are the language of requirements. We will begin by

> ### Rectification of Names
>
> Confucius was asked what he would do if he was a governor. He said he would "rectify the names" to make words correspond to reality. The phrase has now become known as a doctrine of feudal Confucian designations and relationships, behaving accordingly to ensure social harmony. Without such accordance society would essentially crumble and "undertakings would not be completed." (Source Wikipedia https://en.wikipedia.org/wiki/Rectification_of_names)

Figure 25 Confucius (Source: Wikimedia Commons)

clarifying what is a specification and how most specifications are written.

Definition of a specification clause

For our purposes, we define a **specification clause** as a **measure**, a **criterion and an operator**. These terms mean:

- **Measure** is the parameter that is to be a characteristic of the design,
- **Criterion** is the standard value against which the design characteristic is judged, and
- **Operator** is the comparison to be made between the measure and criterion.

> **Specification Clause:**
> - **Measure**,
> - **Criterion and**
> - **Operator**

> Specification Clause:
> **Wait 24 Hours Before Use**
> - **Measure:** Wait Time
> - **Criterion:** 24 Hours
> - **Operator:** ≥ (at least)

The taken measure that meets the criterion in the way dictated by the operator. A specification clause is the lowest unit of a set of requirements and there must be at least one in the set. Because all the specification clauses must be satisfied, people often refer to all the specification clauses together as, the "**Specification**."

We'll first investigate scales because they get progressively more complicated. The types of transformations and operations we can apply to them also change as they increase in complexity.

Transformations and operations are how we write criteria for measures made using a scale.

Metric scales

Level of Measurement or **Scale of Measure** is a classification that describes the nature of information within the values assigned to variables. Psychologist Stanley Smith Stevens developed the best-known classification with four levels, or scales, of measurement:

- nominal,
- ordinal,
- interval, and
- ratio.[6]

This framework of distinguishing levels of measurement originated in psychology but will help explain our need for metrics in specifications.

Each of the metric scales has limitations the define the mechanism of assigning measurements to scales. Only certain mathematical operators can be used on the scales, and only certain statistics are appropriate for them.

Nominal

Nominal Scales assign the measured quantity to a name. The name has no particular order, although it can be a letter or number. What matters is that there is a specific method of determining whether or not the attribution can be assigned the name.

Here are some examples:

- Oak, Pine, Maple, Cypress, Mahogany
- Bill, John, Mary, Joan, Ralph, Linda
- The license plate "number" on your car
- ASME Standard B31.3 - 2016
- USB 2.0
- Annual Car Inspection in Texas
- New Tire Replacement at Acme Tire
- Annual Dental Checkup by Dr. Smith

You can see that names can be applied to a single, characteristic, multiple characteristics, procedures, i.e., anything that can be formally described and named.

Figure 26 Nominal Scales have only categories (Image: Wikimedia Commons).

---

[6] https://en.wikipedia.org/wiki/Level_of_measurement Retrieved July 23, 2018.

Nominal scales only measure classification or membership. The only operators permitted are equal ("=") or unequal ("≠"). For example, here are some criteria using nominal measures:

- All check list items are within tolerable limits
- The material shall be OAK
- Complies with the requirements of ASME B31.3-2016
- The material shall be Class I Lumber
- Successfully Completion of Standard Operating Procedure 20114
- The delivered item is acceptable (meets product definition requirements)

> In Nominal Scales, the only operators permitted are **equal ("=") or unequal ("≠")**.

Note the terms "is", "are", "shall be", "complies with", "successfully completes" are all "=" operators in the sense that characteristics of the measured item equal the defined characteristics of the class.

Ordinal

Ordinal Scales have all the attributes of nominal scales but add order. Examples, include:

- First Born, Second Born, Third Born
- Gold Medal, Silver Medal, Bronze Medal
- Serial Number of your laptop computer
- Step 1, Step 2, Step 3, etc.
- USDA Butter Grades AA, A, and B
- Wheel Lug Tightening Sequence on 2006 Ford F-150
- Earthquake Richter Scale

> Ordinal Scales – Nominal Scales with **items in order**.

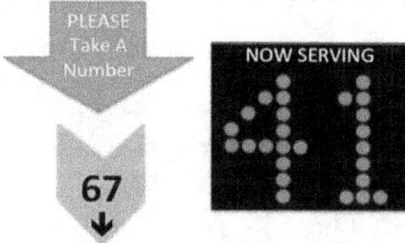

Even though there is a sequence, there is no standard separation (equal intervals) between members in the sequence.

Figure 27 Ordinal Scales have order but no uniform interval (Image: Wikimedia Commons).

Ordinal Scales permit comparison or level. The newly allowable operators are "is greater than" (">") or "is less than" ("<"). Along with the metrics of the nominal scale. For example:

- Hotels rated 4.0 or better by Trip Advisor ≥ 4.0
- Score on the SPED PPD Level I exam >70
- Rated to Withstand up to a Category III Hurricane on the Saffir–Simpson scale
- Recommended for Cancer Stage 2 or Less
- Successful Destructive Testing of every 10th item
- Plywood Grade B or better

> Ordinal Scales add the operators **"is greater than" (">") or "is less than" ("<")**.

> Interval Scales – Ordinal Scales with a **fixed interval**

Figure 28 A hieroglyphic calendar at Elephantine. (Photo: Théodule Devéria – Wikimedia Commons)

> Interval Scales add the operators **"plus" ("+") or "minus" ("-")**.

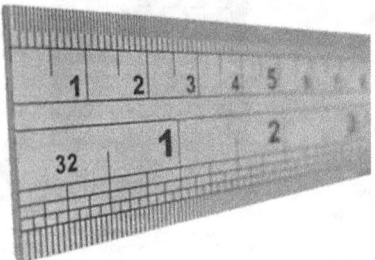

Figure 29. Ruler show standard inches and centimeters. (Image: Wikimedia Commons)

> Ratio Scale - Interval Scale with a **true zero**.

- Heat to red hot color
- Horse race bet to Win (1st), Place (≥2nd) or Show (≥3rd)

Many engineering materials are graded into ordinal scales but the intervals have little or no useful meaning. For example, the difference in properties between Plywood graded B and C is not necessarily the same as between C and D.

Interval

Interval scales have all the characteristics of ordinal scales but have a standard interval separating each item in the sequence. Examples include:

- Temperature in Degrees Fahrenheit
- Julian Date
- Time (unless you count the "Big Bang" theory)

Even though there is a standard interval, there is no true zero. Each measurement, though, is stated with its value and Unit of measure (interval).

Interval Scales permit difference comparisons and important statistical metrics. The newly allowable operators are "plus" ("+") or "minus" ("-"). For example:

- Thermometer accurate ±1.% over a range of 0-212 °F
- Control Setting to hold setpoint ±1.% over one hour
- Timer 0-100 seconds in 0.001 second increments

The intervals are equal but there is no absolute zero.

Ratio

The Ratio Scale is an interval scale with a true zero. Examples include:

- Temperature in Degrees Rankin[7]
- Length in Meters

The ratio scale permits multiplication ("X") and division ("/") operations. This is the most important scale in Engineering, where physical phenomena is scaled up from test results in the lab.

Measurements using alternate systems of units will be proportionate to each other and have the same zero. For example:

- Length 10 inches (254 mm) ±1/8in

---

[7] https://en.wikipedia.org/wiki/Rankine_scale , accessed January 1, 2019.

- Volume 100 cu ft or more
- Yield Stress ≥ 60,000 PSI

Most of the physical parameters we deal with are ratio scales

> Ratio Scales add the operators **multiplication** ("X") and division ("/").

Example: Specifying a TV

Suppose you want to buy a television and you go to your favorite website. How do you narrow your search using some criteria?

Most commercial websites offer a series of selections to help you narrow your search. (See Figure 30.)

Nominal Scale
Brand
Search brands
☐ Samsung (118)
☐ LG (74)
☐ Sony (44)
☐ Sharp (18)
☐ VIZIO (41)
☐ Insignia™ (24)
☐ Toshiba (12)
☐ TCL (16)
Show More

Ordinal Scale
Customer Rating
☐ Top-Rated (3)
☐ ★★★★☆ 4 & Up (3)
☐ ★★★☆☆ 3 & Up (4)
☐ ★★☆☆☆ 2 & Up (5)
☐ ★☆☆☆☆ 1 & Up (5)

Interval Scale
TV Screen Size
☐ 24" and Under (14)
☐ 28" - 34" (36)
☐ 35" - 40" (21)
☐ 41" - 45" (40)
☐ 46" - 49" (36)
☐ 50" - 54" (23)
☐ 55" - 59" (89)
☐ 60" - 64" (8)
☐ 65" - 69" (76)
☐ 70" - 74" (6)
☐ 75" or More (52)

Ratio Scale
Resolution ⓘ
☐ 2160p (4K) (173)
☐ 720p (HD) (31)
☐ 1080p (Full HD) (43)
☐ 4320p (8K) (4)

Figure 30 Typical Selections used to Narrow Product Searches. (Source: Bestbuy.com)

The types of selections offered to you are:

- **Brand**, which are discrete choices. **Brand** is a **Nominal Scale**;
- **Rating Threshold** which is a preference with uncertain intertain intervals between the "number of stars". **Rating Threshold** is an **Ordinal Scale** because the intervals between rating is not standard.
- **Screen Size**, is based on a measured diagonal length inches (a **Ratio Scale**). **Screen Size Range** choices are roughly based on fixed intervals (approximately an **Interval Scale**).
- **Screen Resolution** is approximately a **Ratio Scale** because it has equal intervals and a true zero.

The point is that you are dealing with scales in everyday life.

Criteria

Allowable Mathematical Operators

A specification is a logical clause, I.e., it is either true or false involving a measurement(s) made compared to a measurement(s) **criterion**. The operator and criterion is termed the measurement's "**criteria**". Understanding the measurement scales and the allowed mathematical operations on them is essential in writing specifications. Each specification clause is a measurement made against a measurement (or computed property) standard and a mathematical operator. When the measurement made satisfies the operator on it and the standard, the specification is logically "true". In other words, we say the measurement "meets the criteria."

When is an inch an inch? Compare these four requirements:

1. The lumber must be 1 inch thick
2. The lumber must be nominally one inch thick
3. The lumber must be 1 ±1/8 inch thick
4. A sample from each lot of lumber must be confirmed to be nominally one inch thick.

Number one is vague in that it is virtually impossible to meet an exact numerical dimension. Whatever lumber we choose, it will always be a little off from the exact number.

Number two is better in that lumber is available everywhere milled to the 1 inch nominal thickness according to the American Softwood Lumber Standard, (Voluntary Product Standard, PS 20-15). That standard spells out the allowable dimensions of board nominally 1 inch thick.

Number three is self-contained with a measurement in a ratio scale with a tolerance. Ratio scales can use multiplication and devision (x and /) operators but it can also use all the operators of lessor scales. In this case the tolerance uses the >, < operators of the ordinal scales. Tolerance can be checked with two "**GO/NOGO**" **gauges** (see Figure 32).

Figure 31 Criterion check for carry-on bags.

Number four is an a sampling operation on the measured dimensions of the lumber. Sampling from lot, calculating average, mean, median, etc. are calculated metrics that must be consistent with the measurement scale being used. For example, it's senseless to list the name of every city in New York and try to compute the average city name. It might make sense to compute most frequently used word in a set of city names (probably "city" as in New York City).

Figure 32 GO/NOGO diameter gauge. (Source: Kaboldy Wikimedia Commons)

Measurement Scale Comparisons

Wikipedia published a **comparison** of all four measurement scales[8]. Table 10 shows the type of scale and how they compare on:

Table 10 Comparison of Measurement Scales (Source: Wikipedia)

Incremental progress	Measure property	Mathematical operators	Advanced operations	Central tendency
Nominal	Classification, membership	=, ≠	Grouping	Mode
Ordinal	Comparison, level	>, <	Sorting	Median
Interval	Difference, affinity	+, −	Yardstick	Mean, Deviation
Ratio	Magnitude, amount	×, /	Ratio	Geometric mean, Coefficient of variation

- **Measure property**, which describes what the measurement represents
- **Mathematical operators**, which are can be used for comparisons to set criteria.
- **Advanced operations**, which are permitted on measured values without distorting the values measured.
- **Central tendency**, which are calculated metrics on the values measured[9].

Most specifications encountered will utilize these four scales and the permitted operators and calculated metrics.

---

[8] https://en.wikipedia.org/wiki/Level_of_measurement, accessed December 31, 2018.

[9] https://en.wikipedia.org/wiki/Central_tendency accessed January 1, 2019.

Specifications Defined
Specification Clauses

A requirement "specifies" what is expected from the product. The specification is made up of multiple specification clauses that spell out each mode of performance. Each specification clause is a measure, a criterion and an operator. The criterion is in the same scale and the operator must be appropriate to that scale.

While each clause involves scaled measures, the clause itself is logically true or false (satisfied or not-satisfied). This is an

> While each clause involves scaled measures, **the clause itself is logically true or false (satisfied or not-satisfied).**

Table 11 Sample Specification Clauses from Each Scale

Scale	Measure	Operator	Symbol	Criterion
Nominal	Brand	Must Be	=	Sony
Ordinal	First product from every batch	Must out-perform in testing	>	Minimum Required Performance (of all)
Interval	Storage Temperature	Stays within set temperature by	±	3° F
Ratio	Average Ultimate Strength	Is at least	≥	20,000 PSI

important and unifying feature of clauses and allows them to be combined into a single requirement.

The combination of all the specification clauses is the "**Full Specification**" itself. Usually it is clear whether we are discussing a clause or the full set clauses (specification) so we use them interchangeably.

Normally all the specification clauses of the full specification must be met. It would be tedious, however, to keep writing

Clause 1, and
Clause 2, and
Clause 3, and
……
Clause N.

> **Since the "and" operator is implied in a list, we normally just list the clauses and point out exceptions and contingencies.**

Since the "and" operator is implied in a list, we normally just list the clauses and point out exceptions and contingencies. For example, we don't ask for, "a bicycle **and** the color is green **and** the tires are 26" in diameter **and** the tires are 2" wide **and**…" We just ask for "a green bicycle with 26" x 2" tires."

Things either satisfy the specification or not. When the specification is the design requirement, then a thing is either acceptable or not. There is no "maybe" when the specified requirement is precisely written.

Venn Diagrams

We can illustrate this on something called a **Venn Diagram**. Figure 33 shows a Venn diagram of all possible thing in the **Universe** (U) and the **set of things we define as acceptable, i.e., the Acceptable Set** using our requirement specification.

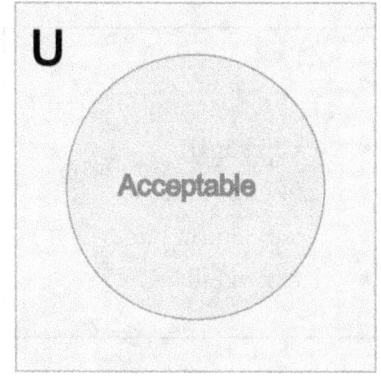

Figure 33 Universe (U) of all things with the set of acceptable things included.

Say, for example, we want transportation using a diesel engine. Of the possible modes of personal transportation (Universal Set), only cars, trucks and Sport Utility Vehicles (SUVs) come with diesel engines. (Not all do but some.) No bicycles or motorcycles do. So, the sets of cars, trucks and SUVs that are diesel powered intersect the set of acceptable vehicles while there is no intersection with bicycle and motorcycles. This is shown in Figure 34.

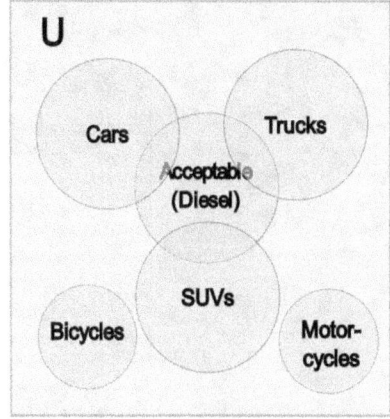

Figure 34 Venn diagram of diesel-powered, personal transportation options.

Venn Diagrams appear everywhere in other forms. Figure 35 shows a diagram used to assess flaws in pipeline welds[10]. The engineer will measure and calculate two parameters and plot a point on the the Failure Assessment Diagram. The point is either in the "Acceptable" or "Not Acceptable" region, i.e., a member of either the acceptable or not acceptable sets.

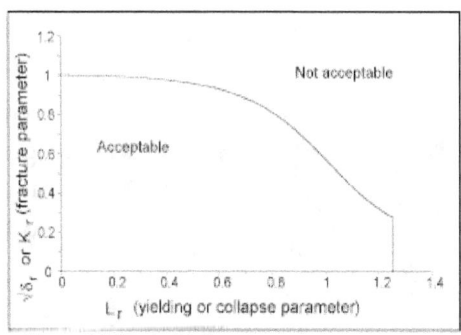

Figure 35 Typical failure assessment diagram for pipeline weld flaws.

Process/Procedure Definitions

You decide to bake a "Sunshine Cake". You go online and download a recipe[11] and buy the ingredients (see Table 12). You move to the kitchen where tools for cooking is stored (measuring spoons/cups, stirrers, bowls, the oven, etc.)

Everything is ready and so, you begin:

---

[10] Cheaitani, Mohamad J, CASE STUDIES ON ECA-BASED FLAW ACCEPTANCE CRITERIA FOR PIPE GIRTH WELDS USING BS 7910:2005, https://www.twi-global.com/technical-knowledge/published-papers/case-studies-on-eca-based-flaw-acceptance-criteria-for-pipe-girth-welds-using-bs-79102005-june-2007 , Accessed January 24, 2019.
[11] https://en.wikibooks.org/wiki/Cookbook:Sunshine_Cake. , accessed 02 December, 2018.

Table 12. Sunshine Cake Ingredients.

- 6 eggs
- 1/3 tsp. cream of tartar
- 1 cup sugar
- 3/4 cup flour
- 1 tsp. lemon juice
- 1 tsp. vanilla

"Separate the eggs. Beat the yolks with a rotary beater until they are thick and lemon-colored. Beat the egg whites until they are foamy, add the cream of tartar, and continue beating until they are dry. Fold the sugar into the egg whites and then fold the yolks into this mixture. Sift the flour several times and add it. Add the lemon juice and vanilla, pour into a sponge-cake pan, and bake."

The last part of the recipe, what to do to make the cake, is the **Process Definition**.

Specifying Processes and Procedures

A **procedure** defines how to implement one or several activities specifies for each step what needs to be done, when, and by whom[12]. Each activity may have a process that transforms inputs into outputs, etc., and, thus, is more akin to a work package. For our purposes, **procedures** are **work packages**.

The simplest definition of a process is a list of steps. These are found in most product instructions. For example, consider the following directions for using Elmer's Carpenter's Wood Glue Interior:

> **Preparation: Temperature, glue and wood must be above 55 degrees F. Surface must be clean, dry and free of oil and grease. Parts must fit snugly. Application: Glue to bare wood only. Spread glue on both surfaces. Clamp 30 minutes. For firmest bond, allow to dry overnight. Cleanup: Wipe up with warm, damp, clean cloth before glue dries. Keep from freezing. Store at room temperature.**

The instructions include preparation, application, cleanup and storage process conditions and steps. If we add, the glue, rags, clamps, temperature controlled drying area, wood pieces and you, we have a procedure (work package).

Requirements Engineering

The formal management of requirements has become a field of its own. Requirements engineering (RE) refers to the process of defining, documenting and maintaining requirements in the engineering design process. It is a common role in systems engineering and software engineering[13]. Requirement management, which is a sub-function of Systems Engineering practices, is also

---

[12] https://en.wikipedia.org/wiki/Procedure_(business) accessed January 13, 2019.
[13] https://en.wikipedia.org/wiki/Requirements_engineering accessed January 11, 2019.

indexed in the INCOSE (International Council on Systems Engineering) manuals.

Chapter Take-Aways

Specifications (specification clauses) are the language of requirements. Each clause has three parts:

1. Measure
2. Criterion
3. Operator

As a clause, it is either "True" or "False". In design, this translates to Acceptable or Not Acceptable.

Problems

1. Assign the following measures to their type (nominal, ordinal, interval, ratio)
    a. Phone Number
    b. Your Grade in School
    c. Age in years
    d. Miss Universe, 1st Runner Up, 2nd Runner Up
    e. Gender
    f. Grade on Last Quiz
    g. Difference in Height Between a mother and father.
    h. Time since you last ate
2. Identify the scale, measure, operator and criterion for each specification below:
    a. You must make a 70% on the quiz to pass
    b. The product must be made in the USA
    c. You must be 16 years or older to get a drivers' license.
    d. No pets allowed inside this building (Except seeing-eye dogs)
    e. Your temperature must be below 100° F to return to school.
    f. Wait 24 hours before use.

The Design Process
Design Process Overview

Designers are usually provided the Clients' functional requirements. Then, they design one and optionally, two things:

1. The production definition that will satisfy the required functional design characteristics, and, optionally,
2. A Work Package that will create the product defined

They have to do these things without running out of resources.

> **Designers must prove the design performs as required.**

Designers must also "prove" that the defined product will perform as required. This assures the quality of the product definition, as satisfying the functional requirements, including any codes, standards and regulations that apply. In short, design is the **creation of product definition data quality assured to meet functional requirements.**

We refer to the defined and proven product or service that satisfies the functional requirements as the **Solution** to the problem posed by the functional requirements. When people say designers "solve problems", this is what they are referring to.

Usually, the designer is tasked to design the **Work Package** that implements the design. The work package is just another designed object, whose function is to realize the design as defined. Our definition of design above still holds but two different things are being designed (the product and the work package that makes it). The functional requirements of the work package is that it implements the defined design within its stated resources.

> **The work package is just another designed object, whose function is to realize the design as defined.**

This dual nature of design process (a product definition and implementing work package) is confusing to most students of design and many practitioners. There are several reasons for this confusion about the design process:

- The end result of design is a defined product and very frequently a work package that implements it. The confusion comes from designing two things at once that influence each other.
- The path taken to the end result can vary based on the knowledge and experience of the designer. The confusion comes from leaps, false starts, dead ends, and other alternate paths that result from a limited view of the possible solutions.

- Satisfying requirements within constraints is more important than optimization. Hence, design management frequently forces the process to converge early to a design with a high probability of remaining feasible to conserve remaining resources. It confuses observers into thinking that all design is deterministic, when it is really just making important early decisions that dramatically reduce uncertainty.
- The **design** of the product definition "task" is NOT the **design** of the work package nor the **execution** of the work package. While both designs consume money and talent, the definition of the solution to the need is separate work from the design of the work package. Neither should be confused with the implementation that executes the work package.

Later, we will discuss some techniques used to design things that are realizable and feasible, i.e., the design process itself.

Proving Performance Using Technology

How is the quality of the product definition assured? By proving that the defined product functions as required using one or more available "Technologies". **Technology** is a product or process for which the performance is known in a functional area we are concerned with. Some of the techniques used to demonstrate functionality are:

> **A Technology is a product or process for which the performance is known in a functional area we are concerned with.**

- Acceptance Testing of the final product or Samples of products
- Engineering Analysis
- Testing of Prototypes
- Testing of Scale Models
- Testing of system components

Some or all of these techniques can be used on in the design of a product.

Technologies known to be capable of proving the design are a "resource" to the design effort. Thus, the more techniques known to apply, the more resources available.

Resources

Real designs are limited by their resources. Resources include:

- **Funds** to design the product and the work package;
- **Technology** that can provide the function(s) required;
- **Skilled manpower** that can identify, specify and prove that technologies work;

- **Tools** (such as software) that can be used to document and prove the design; and
- **Time** to do the design.

It's possible that the design effort will run out of these resource before finishing. A design is **Unfeasible** if it determined that resources will be exhausted before completion.

> **The Design will halt as unfeasible if it exhausts or will exhaust all allocated resources before the design succeeds.**

Design Processes

Analysis and Inversion

Sometimes, if we know what the system is, we can calculate how it might perform. Analysis of the system characteristics and resultant performance can lead to one or more equations that relate characteristics to performance. This is called an equation of the performance function of the system.

Sometimes we can manipulate this equation to give characteristics that perform as required. In these situations, if we know what the system must do, we can calculate what it must be. If we can, the latter calculation is called inversion of the former performance function. This is illustrated in Figure 36.

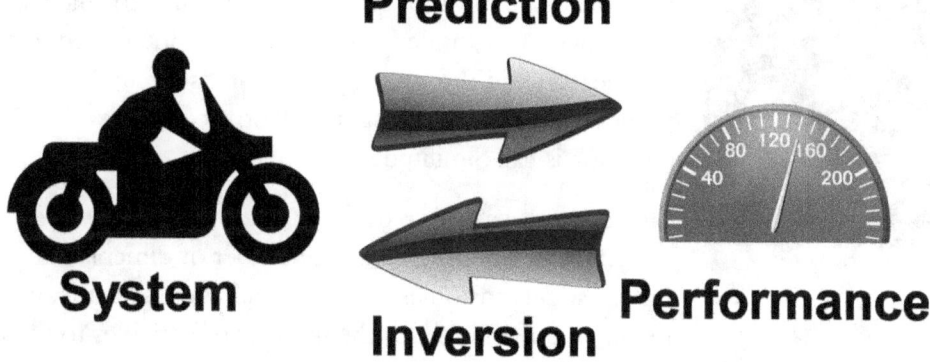

Figure 36 If the function that predicts performance can be inverted, then we can calculate the system characteristics needed from the requirements. This can rarely be done.

Assume we require a performance measure r. If we have a performance function f that calculates a performance (comparable to a prescribed functional performance r) as a function of one defined characteristic d then we could write our criteria as:

r=f(d)

If that function had an inverse f^{-1} we could immediately calculate characteristic d by

d=g(r)= f^{-1}(r)

we say that g(r) is the inverse of f(d). In other words, we have a function that directly computes a characteristic from its required functional performance. This is called inversion.

This is almost never the case. Here's why:

- Each performance measure might be a function of many product characteristics. Generally, there are many more product characteristics than performance measures. Even when a inverting performance measure function can predict a product characteristic, many other characteristics must be "assumed," in advance.
- The system is generally made up of many subcomponents. While the subcomponent functions may be known, their participation in the total system will require a new functional model. Rarely are these system models easy to invert.

When direct inversion of performance functions are not possible, we have to use other methods to specify product characteristics.

Dimensional Analysis, Scale Modeling and Similitude

Perhaps the greatest technical advance for design was the widespread adoption of similitude and scale modeling. Using knowledge of the underlying physics of a phenomenon observed and measured in the lab, engineers can "scale up" the results to predict what full-scale designs will do (see Figure 37). This approach is call Similitude.

Figure 37 12-percent scale model of the US Navy variant of the F-35 (Source: US Air Force)

Dimensional Analysis reduces the number of variables to study in a phenomenon to a minimal number of dimensionless numbers. Dimensionless numbers eliminate systems of measurement and reduce the number of variables to the fewest possible. The lab cannot run its experiments on the fewest test cases.

Engineers then assume the design behavior is similar to the lab phenomenon. This assumption is called similitude. As a result of similitude, the design dimensionless variables predict the design's performance.

The steps are as follows:

- A Dimensional Analysis is done to construct one or more dimensionless numbers that include the independent variables. Dimensionless number are independent of any units of measurement (i.e. imperial units or SI units) but you need enough to capture the most "physically meaningful" effects.

- A phenomenon is studied in the lab until a broad range of dimensionless variable values are covered.
- The lab results are restated in terms of the dimensionless numbers.
- Assuming Similitude, the dimensionless number are calculated for the proposed design and the lab results used to predict the design's performance.

What's most important is that engineers don't need to know the governing **equations** are to scale up: They just need to know what the governing **measures** are and use dimensional analysis to do the right tests on a scale model. Using similitude, the latest discoveries can be applied in designs without waiting for the basic research to explain everything with an exact equation.

Systems Modeling

Practical systems are made up of hundreds and thousands of individual components interacting in complex ways. It would be nearly impossible to check the performance of such systems by hand. The widespread availability of low cost computers and capable software allows us to predict system performance in most modes of interest.

For example, refinery piping might be checked for:

- Hydraulic Flow;
- Pipe Stresses:
- Distortion/Leakage at Flanges (between piping and at equipment) and stresses in flange bolts
- Expansion and contraction during hot and cold states
- Vibration during operation and during storms and earthquakes
- Corrosion and deterioration from fluids inside the pipe
- Corrosion outside the pipe
- Thermal losses from the fluid inside to the environment
- Effects of welding on underlying metallurgy

Figure 38 CSIRO Computer model image of a rogue wave smashing into a semisubmersible platform. (Source: Wikimedia Commons)

This checking might be done by several software packages and/or modules withing the same package. Piping Engineers, Materials/Metallurgical Specialists, Civil Engineers, etc., are used to oversee the data entered into these simulation packages and interpret the results. It would be almost unthinkable to analyze most piping systems by hand for all but the smallest plants.

The field of managing complex systems is called Systems Engineering[14]. Although systems engineering is mainly applied at the requirements vs function level, most engineering disciplines model the technical performance of subsystems and try to roll up low level performance to higher level requirements. Imagine trying to use the flow rate of a pump and response time of a control valve to predict the quality of the gasoline coming out at a refinery!

This field of systems modeling uses engineering software to predict high level technical performance. Since several types of performance are required, several engineering software packages might be used. The same component might be modeled in multiple packages, depending on their functions within the design.

Iteration

If we have a way to calculate how a system might perform, we can just keep making changes to it until all the requirements are satisfied. This is call iteration. Inversion is never necessary, but we have to be really good at guessing.

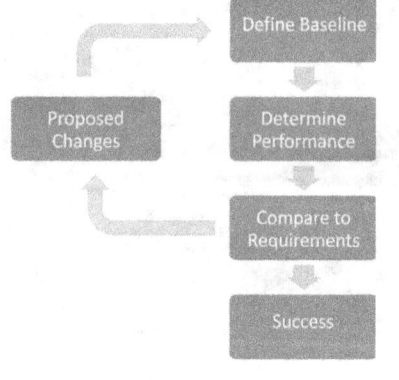

Figure 39 Iteration Process

The process of iteration is shown in Figure 39. A design is approved to be the "baseline". The baseline is the working design from which performance is checked and to which all changes will be proposed. The performance of the baseline design is determined and compared to the requirements. If the requirements are satisfied, the design is declared to be a success and halts. If not satisfactory, changes are proposed. If the proposed changes are approved, they become part of the new baseline.

Every iteration can be thought of as an experiment with an outcome. In the simplified model of Figure 6 every iteration can only end in success or another iteration. If an acceptable design is never found, the design can continue forever. Of course, this is never done in practice because you run out of either time of money, two essential resources.

---

[14] https://en.wikipedia.org/wiki/Systems_engineering accessed January 11, 2019.

Refinement and Decomposition

H. Eshuis gives the following definitions of refinement, abstraction and decomposition[15]:

> ...*The addition of detail to a design, or in other words, making a design more concrete. Refinement of a design at a high level results in a design at a lower level, that is, a design with more detail. Addition of details is necessary because not all aspects can be taken into account in the first design. In a later design, the remaining aspects are also taken care of, and then more details are added to the design.*
>
> *Refinement is the counterpart of abstraction. Abstracting a design means leaving out irrelevant details of that design. In that case a higher level design is obtained. By refining a design, on the other hand, a lower level design is obtained, that has more detail than the original design.*
>
> *Refinement also interfaces with decomposition. Decomposition is used in the context of systems. A system can be decomposed in subsystems, that together form the complete*

Refinement increases detail. Abstraction removes detail.

Decomposition subdivides. Unification combines.

---

[15] H. Eshuis, Refinement in object-oriented analysis and design. University of Twente, Faculty of Computer Science, Information Systems, Wierden, August 1998.
https://pdfs.semanticscholar.org/b5b3/045dee401a14025563ed95424fd13e7deef3.pdf retrieved 14 Nov 2018.

Every design "decision" to set a baseline constrains the systems around it and inside of it. Hence, if a car designer decides on a diesel engine, the surrounding fuel and control systems are constrained and the future choices of engines (i.e., to diesel) are also constrained.

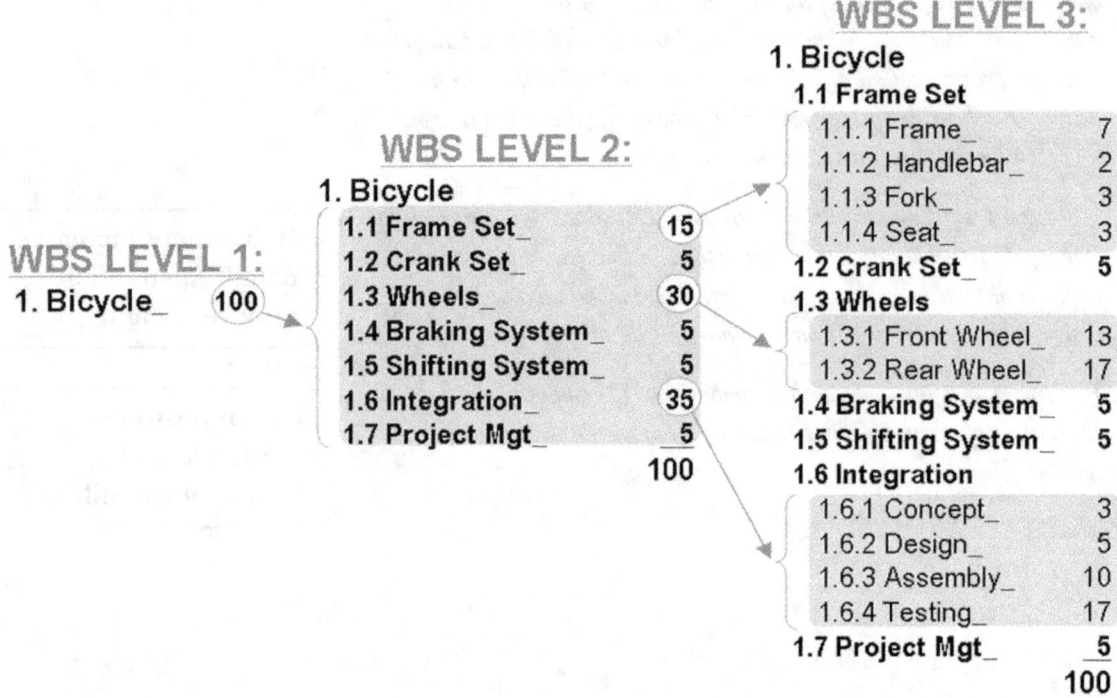

Figure 40 Decomposition of a bicycle design WBS along its constituent parts. Note that the design has not become refined to more detail. (Source: Garry L. Booker, Wikipedia).

Refinement and Decomposition are essential design processes in all designs. The process must meet the functional requirements of the overall system, while the requirements of itself and the surrounding systems are constrained to be more specific.

Writing to the appropriate level of detail is always an issue with requirements. Writing to a very abstract level permits many possible design choices, resulting in many possible refined or decomposed requirements within it. While staying abstract (leaving your options open) sounds good, it delays the labor-intensive detailed design work and can impact the schedule. Conversely, retracting a design refinement later because it is found to be impossible or unfeasible, also delays the schedule. The usual course is to converge quickly to a concrete design with a high probability of success.

Standardization of Components
If is very efficient to use standard components whenever possible. They offer many advantages over non-standard or specialty components.

Standardized Components

Standardized components are functionally defined by an accepted standard. The components they offer many advantages over specialty components:

- They have functions, features and characteristics defined by 3rd parities, such as governments, standardization boards and professional societies
- There are frequently many suppliers who compete on price, delivery and additional features and performance not required by the standard.
- Other engineering standards frequently "pre-approve" such components in the scope of their designs as having acceptable performance.
- The designation of such components are frequently only a matter of citing the standard description.

Specialty Components

Specialty components offer performance not completely covered by a standard. Specialty components offer many advantages over standard components:

- Their performance is in areas not yet covered by a standard.
- They offer an entirely new and often superior level of performance to standardized products
- They fall under the standard only by additional analysis

Designers try to use standardized components wherever possible because they are already acceptable under most performance checks. When specialty items are used, designers have to document their source and justification.

Preference for COTS Products

There is a deliberate bias among designers to converge to a Commercial-Off-The-Shelf (COTS) product over designing it from scratch. Often suppliers will publish the rated performance of their products as a set of specifications (specs). These are minimal values of stated measures that can be relied on for comparison to required specifications. If their specs meet or exceed the required specs, designers prefer to buy product, rather than build them (see Figure 41).

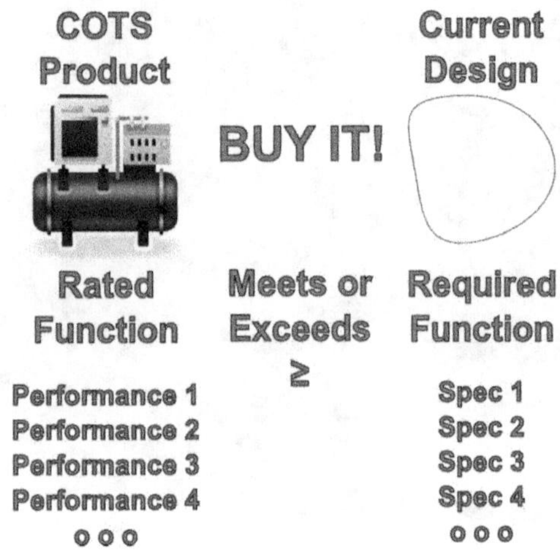

Subsystem product specs are used to prove system performance but it is now the responsibility of the product manufacturer to assure this performance. They normally place many additional conditions on the design as a result. These conditions now become part of the requirements of the system.

Engineers and other designers are not relieved of their responsibilities to confirm performance by selecting a standardized product. The recommendations of a credible supplier, with documented performance data, makes the prediction of the overall system performance much easier.

Even if supplied parts and equipment create new requirements, their cost and available are usually too attractive not to use them. To avoid confusion, we try to focus on proven product performance, rather than just rated performance specified by suppliers and vendors.

Phasing

While design might proceed directly to its final, implementable result, in a practical world it is constrained by the need to manage it in such

Figure 41 Designers prefer COTS products when their rated function meets or exceeds the required function.

a way to economize its use of resources. The most common management process is phasing. One example is given in Figure 42.

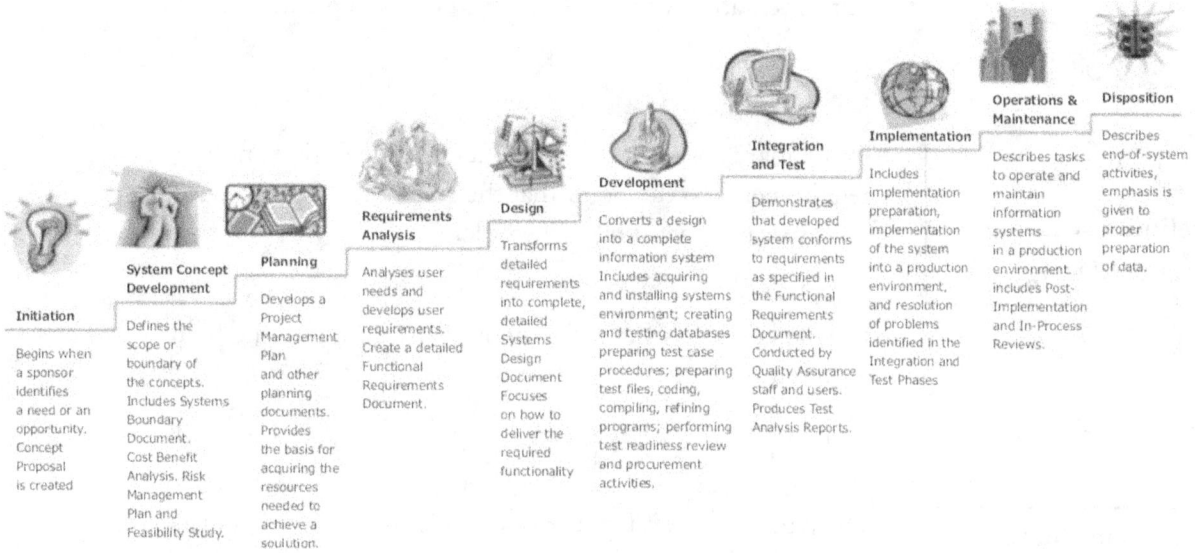

Figure 42 Systems Development Life Cycle. (Source: US Department of Justice, Wikimedia Commons)

By breaking up the design work into several phases, the remaining requirements on resources can be estimated and the business case can be re-evaluated. This can curtail costly overruns and highlight other problems. In the following sections, we'll highlight some common design phases.

Pre-Design Processes

Before design begins, several things are done to put in place the elements needed. This includes such things as defining the requirement, identifying the resources available to the design effort (funds, staff, schedule, etc.) and what phases will be used to manage the design process.

Pre-Design Studies:
- **Business Case**
- **Feasibility of Design**
- **Feasibility of Implementation**

Business case

Business cases predict a return or benefit on the resources allocated for an implemented design. Most commercial designs are required to fulfil a specific business case.

Design Feasibility

Design Feasibility reviews are a formal study of an abstract design to determine if the requirements can be satisfied before consuming allocated resources. It also considers the risk that the design will be "Impossible," i.e., that the requirements cannot be satisfied without violating a physical or regulatory law.

Implementation Feasibility

Implementation considerations should be part of pre-design study or of the design itself. It's quite possible that implementation of the will be given only cursory consideration before beginning or the full implementation will also be "designed" along with the solution.

Design Phases

Preliminary design

Preliminary design is the refinement of the design and the requirements to the point that the functional performance of the systems can be verified as a whole.

Detailed design

Detailed design is the refinement of the design to the point it can be acquired or built.

Implementation Phases

Subsystem Fabrication/Manufacturing/Acquisition

The parts of the system are purchased, manufactured or fabricated before the full system is assembled.

System assembly

The various parts are assembled into a functional system. When completed the implemented (as built) performance can be tested.

Startup/Acceptance tests

Rarely do the owners of a new large system just "throw the switch," and start full operations. Most designs have a formal program of acceptance tests that assure all systems function as required.

Operation & Maintenance

Most returns are from plants providing goods and services. Once the plants are implemented and operational, their Capital Expenditures (CAPEX) are depreciated or amortized over time. The depreciation and amortization expenses are considered, along with Operations and Maintenance (OPEX) expenses to calculate profit.

Sometimes, more CAPEX will lower OPEX, as better built plants will have lower downtime, more productive uptime and longer lasting equipment.

Design Phases
- **Preliminary Design**
- **Detailed Design**

Implementation Phases
- **Make or Acquire Parts**
- **System Assembly**
- **Startup and Acceptance Tests**

Shutdown & Remediation

At the end of a plants lifecycle, there is a final cost to shut the plant down and remediate the site. Shutdown and rememdiation can be a significant cost because of contamination and other issues with the old equipment, soil and groundwater underneath.

Baselining

Each approved product configuration is known as the **Baseline** design. After a baseline is approved, subsequent work extends from the baseline configuration. As design choices are made, the set of remaining possible candidates grows smaller and smaller. As design choices are officially "accepted" as baseline, there are fewer and fewer choices remaining to be made until a final, acceptable design is defined. We say that the design "converges" to an acceptable solution (see Figure 43).

Most large designs begin with a document called a **"Design Basis"** or similar name. The design basis contains all the information considered baseline at the initiation of design. It is the "first" baseline of this phase of design.

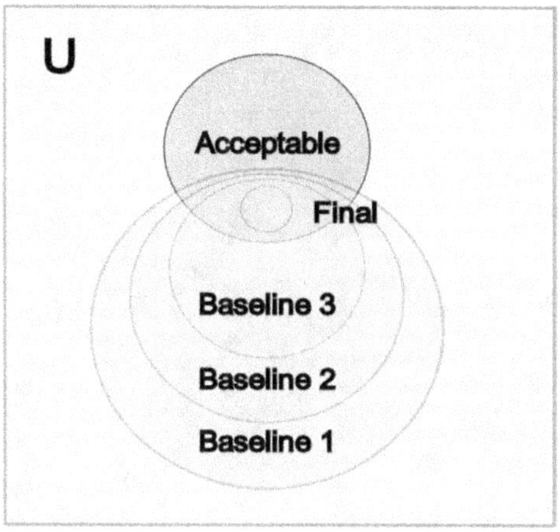

The design basis includes:

- Drawings, tables and charts to be used in the design;
- Codes, standards, procedures to be observed or followed during the design;
- Resources available for the design;
- Constraining deadlines, required reports, liaisons to permit customer oversight.

Figure 43 Successive baselines converge to a fully acceptable design.

The design basis is a very important document and the designer must recognize and extend it in his subsequent work.

Chapter Take-Aways

Design is the definition of proven products and services that will satisfy functional requirements. Optionally, the designer may also design the implementation of the solution.

Designers use several techniques to define products and prove they meet functional requirements:

- Invert functions that predict performance to compute values of product characteristics;
- Use similitude to prove performance on scale models or scale up laboratory tests.

- Use systems software to predict behavior of combined components
- Iterate successive sets of product characteristics until performance is achieved.
- Refine a product by fleshing out its detail for analysis
- Decompose a product into multiple, interacting subsystems, then addressing each one individually and n combination.
- Prefer standardized components and COTS products, where available
- Phase the design to recheck requirements at increasing detail with rollup to the business case.
- Baselining to insure all designers work from approved data and performance is converging to requirements.

Designers use these techniques as needed during design.

Problems

1. You buy a car that is rated at 17 miles per gallon in the city, 20 miles per gallon on the highway and has a 20-gallon gas tank. Is this the right car for a trip in an area where gas stations are 450 miles apart? Why or why not? What predicted performance is used to prove or not prove it meets the acceptable criterion?
2. Describe what might be considered before applying laboratory test performance to a real-world situation.
3. In a system of components, why might the performance of one component be affected by the performance of another?
4. Is the adage, "if at first you don't succeed, try and try again," an example of iteration?
5. Is choosing the members of a baseball team an example of refinement or decomposition?
6. Is choosing the bath type (tub, shower, both) then the hardware (tile, faucets, lights, shelves, etc.) refinement or decomposition?
7. Describe the last Commercial-Off-The-Shelf product you bought for a purpose. Did it satisfy your needs?
8. Describe how cooking a meal actually has phases from designing the meal to washing the dishes.
9. If a football team uses scouts to find players to improve the team, how might the scouting strategy change after each new player is added? Is this an example of baselining.

Defining Things That Work

Introduction

In this chapter we will explain the output of all the designers' work. After all the design processes are executed, what should be in the report?

By the time design begins there is a:

Design Basis – Everything the designer should assume is decided about the design to date.

- Requirements as currently understood
- First Baseline of the Design
- Any constraints on Technologies that can be used in the design

Scope of Work – The level of detail of design to be accomplished and of the implementation work package, if included.

Resources Budget – The resources the designer can use to complete the design

- Monetary Budget
- Schedule and Milestones

Management Plan – How the designer will show progress and deliver final results

- Required Reports and Deliverables
- Customer Liaisons
- Approval Procedures

> Most Designs start with:
> - **Design Basis**
> - **Scope of Work**
> - **Resources Budget**
> - **Management Plan**

In the simplified life of the classroom, much of these formal details will be reduced to a few representative documents for problems designed to be completed in a semester.

Defining Products

Choice of Representation

Representations are displays for entry and review. They can be drawings, narratives, tables, specialized forms and other documents. The representation(s) used to document the design have to be capable of showing all the parameters that define it for analysis and for implementation. (Stated or not, the requirement that the design is implementable is always present.)

> **Choose representation(s) capable of showing all the parameters that define it for analysis and for implementation.**

The following types of documents are normally required:

- Definitive layout – A drawing or model that shows everything all put together;
- Essential Design Details – An explanation of how the design is assembled from its components and material;
- Detailed component designs – Parts that must be made (not purchased)
- Standards Parts list and Specialty item list
- Special Implementation Instructions for Transport, Assembly, Manufacturing and Operation

The most common representation is the "**drawing**".

Anything that shows data in visual form is a **drawing**. Engineers and designers have evolved a system of representing data in manual drawings using descriptive geometry. **Descriptive Geometry** is the branch of geometry which allows the representation of three-dimensional objects in two dimensions by using a specific set of procedures[16]. Using the techniques of descriptive geometry, important details of the design can be determined in views showing their measurements precisely. Drafters are specialists in applying these techniques to determine and show these measurements.

The first design computer tools emulated descriptive geometry techniques (**Computer-Aided Drafting**). Gradually, the importance of representing underlying data in full detail was recognized. From underlying data many views of the same data could be generated or rendered. Today, most **Computer-Aided-Design** (**CAD**) systems concentrate on capturing and storing the data as needed by the application. Drawings or other representations are generated for data entry/checking and to furnish it end users in their preferred format.

Essential Parameters

Every parameter used in proving functionality or implement-ability is an **Essential Parameter** of the design. If a 10" diameter shaft was used in a calculation, the critical 10" diameter must be shown in the design documentation. Every essential parameter must be documented in the design so that it can be implemented.

Every part that is assembled to provide an essential parameter must also be shown with instructions on assembling it into the whole. If a 10 foot thing is made from two 5 foot things, then you have to show how to assemble it. Then, you have to list how to make or buy the two 5-foot things.

Figure 44 Drafter working in engineering department, 1959 (Source: Seattle Municipal Archives)

Engineers and designers have evolved a system of representing data in manual drawings using descriptive geometry

Drawings or other representations are generated for data entry/checking and to furnish it end users in their preferred format.

Essential parameters are used in proving functionality or needed for implementation.

---

[16] https://en.wikipedia.org/wiki/Descriptive_geometry. Accessed January 19, 2019.

When we buy a bicycle it comes disassembled in a box. We get instructions with diagrams that show:

- How to identify the individual parts,
- What order to assemble them in,
- What part gets put where, and
- How to check if the assembly is correct, including any measurements to make.

The design of the bike would not have been complete without the assembly instructions.

Data of Record

Most documents repeat data for association with other data or for extra convenience and clarity. Anyone who has filled out paperwork in a doctor's office knows how many times the same data is repeated on form after form after form.

Most designs have an internal structure when there is a "most authoritative" location of a data item. The **Data of Record** is the location where the authoritative data value is stored. All other uses of that data look to that location for the correct value.

In the US, the first data of record for YOUR NAME and DATE OF BIRTH is your Birth Certificate. Birth Certificates are filed with the clerk of the county or other records clerk. Most additional forms of identification, e.g., driver's license, passport, etc., ask for a certified copy of the birth certificate to insure that the spelling and date of birth are correctly copied. The clerk issues such certified copies.

> **Your Data of Record**
> **At Birth**:
> **Birth Certificate** - YOUR NAME and DATE OF BIRTH
>
> **After Marriage, Divorce, etc.**
> **Birth Certificate** - DATE OF BIRTH
> **Court Document/Order** - YOUR NAME

To change your name, you have to go to court and get a legal document. Name changes are due to marriage, divorce, or other reasons. The order changing the name is filed with the clerk of the court, county or other jurisdiction. The new order is the new data of record for YOUR NAME. The birth certificate continues as the data of record for your DATE OF BIRTH.

Every item of data in the design will have its own data or record. If you want to change the data YOU MUST change the data of record and update all the other representations. As you can imagine, updating data in all its locations is very tedious but very necessary in design.

Rendering of Views and Reports

One answer to the data update problem is to store it in as few locations as possible and use computers to update all the uses. If a view of the data is completely determined by other data, that view can be "generated" as necessary. **Rendering** is the generation of a view or report completely from existing data.

> **Rendering is the generation of a view or report completely from existing data.**

The most common use of rendering is the creation of a 3D rendered view from geometric data defined in a CAD system. Figure 45 shows an example.

With todays modern CAD system many different types of views, drawings and reports can be generated from the data defining the design. This is "rendering" in the general sense. Such renderings are used to assure the data is correct by viewing it in a format allowing it be easily checked.

Ultimately, the "on-screen", live view of data that most CAD applications provide is a rendering of the data stored within. In the years since CAD was introduced, the quality of the representations onscreen has increased dramatically with the power of the computer and the quality of the displays. These improvements included:

- Wire Frame
- Hidden Line Removal
- Real Time Refreshed Views (vs manual)
- Color
- Shading
- Panning, zooming, rotation
- Snapping to points on the Design
- Perspective
- 3D (Dual views for Eyes)
- Hologram generation
- Control by Head Movement (VR)

Each improvement increased the dimensionality of the display and, thus, the immersive effect it had on the designer. The more immersive the display, the easier for the designer to check and assure the design.

Figure 45 Rendering of 3D Shaded View of a Carburetor. (Source:

Figure 46 User interacts with Design using Virtual Reality. (Source: US Air Force)

Anticipating Implementation in Design

The unstated requirement that implementing the design MUST be possible and feasible means that designers must check for this requirement. This is done by:

- Avoiding features that are difficult or impossible to build
- Preferring features that are cheap to build or fabricate
- Adding sufficient detail such that the created items can be built
- Avoiding excessive details that limit the options for manufacturing or fabrication

> **You need enough implementation detail to get what you need without driving up the cost with unnecessary constraints**

It's a situation where you need enough detail to get what you need without driving up the cost with unnecessary constraints.

Frequently, the designer has a manufacturing or fabrication facility in mind while designing. In that case, choices can be made that lend themselves to shop processes currently available to implement the design. Preferred fabricators and manufactures can be brought in to review the design and suggest ways to reduce costs.

> **Frequently, the designer has a manufacturing or fabrication facility in mind while designing.**

Proving Function

Every requirement of the design must be met for success. It's the job of the designer to set essential parameters and use them to prove each requirement has been met.

Analysis

Analysis is the process of breaking a complex topic or substance into smaller parts in order to gain a better understanding of it[17]. Our analysis is generally guided by the nature of the requirements and the features of the design we want to set to meet them.

For example, if our design required:

The weight shall not exceed 10lbs,

Then we'll need an analysis of the design to determine its total weight. If our design consists of four parts and some miscellaneous materials, we can analyze each for its own weight. We:

- find the weight of all the parts and of the miscellaneous materials,
- add them up.

---

[17] https://en.wikipedia.org/wiki/Analysis, accessed January 20, 2019.

Table 13 Analysis of Design Weights to Determine Satisfaction of Weight Specification.

Component	Weight (lbs)	
Part A	2	
Part B	1	
Part C	3	
Part D	3	
Misc Materials	1	
Total	10	✓

Our analysis is done. We compare the total weight measure (10 lbs) to the criterion (10 lbs) and the operator (do not exceed) and **find the specification satisfied**.

The source of those weights, our analysis and our rollup to a specification is part of the "proof" of the design. In most design documentation, the analysis, rollup and requirement test is so extensive that it is usually just called the "Analysis" section of the documentation.

Rollup to Requirements

As the design gets more detailed, the impact of low-level feature changes can spread to more high-level requirements. The designer must determine, through his analysis, how a low-level feature impacts high level requirement(s). This is termed "**Rollup**."

> **Rollup is determining the impact on high-level requirements of low-level design changes.**

Propagation of Constraint

Every design choice implies new constraints on the design. The new constraints must be considered wherever their impact is felt. The incorporation of design choice consequences into all other parts of the design is called **Propagation of Constraint**.

For example, you get the layout of the garage just like you want it: You can get to everything and its looks organized. Then you decide to buy a new car. The car is longer and slightly wider. Because of the change in length and width, you have to rearrange the garage so that the new car fits. The constraints of the new car have been propagated through to your garage layout.

> **Propagation of constraint is the incorporation of design choice consequences into all other parts of the design.**

Constraints on Implementation

It is very common for a design choice to impose **Constraints on Implementation**. Unless these constraints render the design impossible or unfeasible, they are accumulated as requirements on implementation. The design analysis only satisfies functional requirements. The implementation analysis will satisfy these new requirements on implementation on implementation.

Common implementation requirements are for:

- Procurement, Fabrication, Manufacturing, Transport, Assembly and Acceptance;
- Tooling and Training for Operation and Maintenance
- Special Shutdown, Disposal and Site Remediation

> **Design choices impose constraints on implementation.**

These constraints are documented as implementation requirements.

Technologies used

Technology is a product or process where the performance is known. In cases where a technology is used in the design, the known performance is used in part or all of the analysis of the design.

The simplest case is the purchase of a **COTS** product with rated performance. Rated performance means the manufacturer has designed the product to meet or exceed the rated performance under the stated operating conditions. The designer relies on the rated performance in his analysis of the system in which the product is being used.

For example, one requirement of the design's subsystem is to pump water at 60°F at 12 Gallons per Minute (GPM) past a load of 20 Pounds per Square Inch (PSI). You do the following:

- Find a product rated at 15 GPM past a load of 20 PSI.
- Select the product and incorporate the product's hydraulic performance (called pump curves) and other performances and new requirements into your system models.
- Recheck the hydraulic model with the actual pump (it works) and check the other consequences as well (size, power requirements, nozzle loads, etc.). Using the new pump's performance and consequential requirements, the predicted performance from all of the other analyses is still acceptable.
- Ask for and get approval to buy the pump.
- Add the pump model and rated performance into the essential parameters of the design.

Figure 47 Prakash 1.5 HP centrifugal pump. (Source: Wikimedia Commons Prakash Worldwide Co. Inc)

This example illustrates how the design detailing and the analysis of its performance go hand in hand.

Assurance

Rarely is a design analysis accepted just because the designer, says so. There are several checks used to assure that designers analysis is correct and its consequences understood.

Certified and Licensed Skills

Once way of assuring design analyses is by certifying the skills of the designer/analyst. This is done by considering the resume of previous experience, personal recommendations and other endorsements.

Designer skills can be certified by third (3rd) parties like colleges, universities and professional societies. The degree or certificate is offered by organizations with a reputation for high levels of quality

> **Certifying the skills of the designer/analyst.**

and integrity. A certification program might be governed by a standard that spells out it's required skill set(s).

For critical skills where public safety is at stake, designers are often licensed by governmental bodies. Licensing usually applies to:

- Engineers
- Surveyors
- Geologists
- Doctors
- Nurses, etc.

Most professions making judgements and choices for the public require licensing.

Procedures

Designs also select proven design procedures to analyze their designs. **Design Procedures** are steps leading to designs with proven performance. These procedures are found under many names:

- Guidebooks
- Handbooks
- Standard Procedures
- Published Methods
- Codes/Standars
- Recommended Practices

Most designers are delighted to have a procedure available to guide their analysis.

Designers are rarely absolved of the responsibility to conduct their own analysis. Even if they rely on a published or standard procedure, they must decide it applies to the situation at hand. They are always responsible for the correctness of the procedure itself.

Checklists

Checklists are another method of assuring the quality of a design and its analysis. Checklists list:

- All the data elements normally found on a document;
- The data that MUST be found on the documents If the document contains the date of record;
- Typical data ranges and units;
- Important cross checks with other documents; and
- They also list common mistakes designers make.

Keep in mind that the designer who developed the checklist has probably made all those mistakes themselves.

Approvals

Approvals are consents a formal requests that a choice or change be accepted as baseline. The final check on the design is going to the client for approval. Approvals generally circulate to all affected parties in both design organizations and their clients, i.e., all affected parties. So, the design change request usually contains the analysis that justifies it.

Once a change is approved it becomes part of the baseline design. The baseline design is then used for all subsequent work.

Defining Implementation

The full development of implementation planning is outside the scope of this book. It is useful to outline some of the considerations that designers take into account while designing.

Essential Processes/Procedures

Design often anticipates that certain processes and procedures will be used for implementation. Essential implementation procedures are those that were required in the analyses proving the design. If these implementation procedures are NOT used, then the analyses are in question.

For example, the stress analysis of pipe might presume that all weld joints were heat treated. If these joints are not heat treated, then the stress analysis in no longer valid.

In another example, a steel mill is designed where a trained operator runs the loading and unloading of steel and operates the mill rollers. If the operator is not trained, then the analysis of the safe and efficient operation of the mill is in question.

> **All essential processes and procedures must be made part of the requirements for implementation.**

All essential processes and procedures must be made part of the requirements for implementation.

Preference for Standard Procedures

If a standard implementation procedure is available, then it is preferable to use it. In the above two examples, standard heat treatments and standard training courses might be available to use in the design analyses and specify in the implementation requirements.

Products and Material Needed and by what process

Implementation might also require specific products and materials. These required products and materials are now required of the implementation.

Resources Needed to Implement

Other resources might be needed because of design features. Implementation design will include these resources in the work package.

For example, if the design implementation includes the erection of a 100 foot vessel, then a crane that can perform that lift must be available for use on that day. A crew must be available to plan and execute the lift and secure the vessel. Although these resources might not be part of the design, they are part of an implementation work package.

Assurance

Inspections and acceptance tests are normally part of the work package design. Acceptance testing of the total system is generally considered a separate phase of the implementation.

As-Built Updates

Once the system is built, accepted and operational, its documentation should be updated to reflect its final configuration, or, "**As-Built**." As built documentation is re-issued after the updates are completed.

As-Built documentation is updated to reflect the final configuration of the implemented design

Modern designs maintain documentation for the life of the product or service. Frequently maintained on the computer, some are calling this the "**Digital Twin**" to the product or service, depending on how accurately the documentation is maintained.

Reporting Designs

Parts of the Final Report

Design Starting Point Recap

This should be a summary of the starting point of the design with reference to all the documents furnished to or agreed between the designer and client. These include specific documentation on the:

- Design Basis
- Scope of Work
- Resources Budget

Any amendments should also be listed

Final Design (As Built)

The documentation of the final design should be described with major documents listed or referenced. The objective is the clear understanding of the design such that the subsequent analysis should be understood.

Analysis

The analyses showing how the design satisfies each of the functional requirements of the design should be given or referenced if reported separately. There should be no unaddressed and no unsatisfied functional requirements after this section (unless the work was terminated prematurely).

Resources Used or Consumed

A short summary of the resources consumed during the reported work. It should report:

- Costs
- Schedule
- Customer-Furnished Materials

Often, this is reported in a second volume, not meant for general reading.

Essential Facilities and Tooling

Specialized facilities and skills that were used is defining and proving the product should be reported. If the work is ever extended, these resources will be needed again.

Critical Technologies and Operating Principles

If not documented in the final design, technologies and operating principles that need to be used in the implementation should be described

Optional Parts of the Final Report

Project successes, lessons learned

It is a good idea to document the successes and failures of the design somewhere. The goal is to not repeat the mistakes of the past.

For legal reasons related to patents, it should never be reported that an approach has been abandoned. Just report that another approach was chosen.

The client has paid tuition for every lesson learned. They may prefer this section be reported confidentially.

Suggestions for Future Work

There is never enough time to pursue all of the alternatives to the final solution. Most designs end with many possibilities to extend or improve the work.

> "We didn't lose the game; we just ran out of time."
> Vince Lombardi,
> American Football Coach.

Now that the designer is familiar with the need and potential solutions, it is a good idea to explain how they would continue with more resources. It could be the beginnings of your next job!

Chapter Take-Aways

No matter what design processes are used, two items have to exist at the end of design. These two items are:

- A definition of the product or service, and
- An analysis showing each Functional Requirement was met.

We include these items in a "final report", to explain the work. Other parts are usually included to aid the reader in understanding the work and how it was completed.

A representation of the product must be chosen that shows every:

- Essential parameter; i.e. those used in the analysis and
- Essential Process, i.e., those presumed by the analysis to be used during the implementation.

Essential parameters and processes are, in turn, imposed on the implementation.

Certain representations will contain the data of record for each essential parameters and processes. All others should be rendered from it or cross checked against it.

Work packages are another object to be designed to implement the product or service. Good designs will be easily implemented, and all designs should anticipate it. Implementation will meet its own requirements, including those imposed by the design.

Problems

1. Compare a design that starts as a "clean sheet of paper" to one which starts with a baseline.
2. What is the main requirement of design representation? Can you see it might not always be a drawing? It could be a data sheet with blanks filled in or a part number with rated performance.
3. Is every analysis, an engineering analysis? While engineering analysis uses scientific principles, give an example of analysis that applies other principles.
4. Why are basis, design representation and analysis three topics in every design report?
5. Describe the last checklist you completed. What was being checked by the checklist?

Design Disciplines & Technologies

Overview

Engineers specialize in applied science or technologies that already use them. Applied science takes known scientific principles discovered in the laboratory and scale them up to to predict design performance. Scientific principles are often stated as a mathematical model of the physical work, so mathematical tools are studied to help apply science to the real world. We also have many technologies that have applied science and mathematics to give designers predictable performance.

There is great support for Science, Technology, Engineering and Mathematics (**STEM**) as a future career for young people. Table 14 compares and contrasts the common STEM areas for their contributions to design. These and other areas contribute to designs of many types.

However, scientific principles are not the ones that can be applied to prove a design meets a particular requirement. Some functions are non-scientific in character or need no scientific principles to prove their satisfaction. The person applying the principle has to be sure the principle applies to the design and correctly predicts performance.

For example, requirements issued to plumbers, electricians, carpenter and other trades require little science to satisfy. Their "designs" are done at the jobsite, with a final implementation as a result, with

Figure 48 Designers are engineers, technicians and other professionals who apply principles to prove

Table 14 Comparison of STEM Areas by Contribution to Design

STEM Area	Contribution	Examples
Science	Predictions of physical and biological behavior	• Physicist • Chemist • Biologist • Geologist
Technology	Products and Processes with Known Performance	• COTS Products • Standard Operating Procedures • Manufacturing Methods
Engineering	Specialists in Proving Performance using Scientifc Principles and Technologies that Use Them.	• Civil Engineering • Mechanical Engineering • Chemical Engineering
Mathematics	Tools for Expressing Scientific and Other Principles in Useful Formats	• Algebra • Geometry • Calculus • Statistics

paperwork documenting all required steps. Trades construct in ways that meet their drawn and written requirements, while obeying building and municipal codes, verified by inspection. They are applying principles that need no science to prove they have satisfied their stated requirements. Their documentation and inspector's approval demonstrate satisfaction of requirements.

In this chapter, we'll review a few popular engineering disciplines to highlight the products and services they design and the major principles and techniques they apply to prove them. Then we'll highlight a few more types of disciplines that also design or contribute to it.

Engineering disciplines

As you might imagine, for every scientific discipline or combination of disciplines, there is an engineer specializing in applying it. Frequently, these application methods are combined into an engineering discipline of its own. If it's popular, college and universities will offer courses of study in the most used scientific principles and mathematical tools.

> "Any idiot can build a bridge that stands, but it takes an engineer to build a bridge that barely stands."
> - Anonymous

Civil Engineers

Civil applications are generally soil-based structures that shape the physical environment. Civil engineering is a professional engineering discipline that deals with the design, construction, and maintenance of the physical and naturally built environment, including works such as roads, bridges, canals, dams, airports, water and wastewater systems, pipelines, and railways.

Civil engineers conceive, design, build, supervise, operate, construct, and maintain infrastructure projects and systems in the public and private sector[18].

Civil Engineers work under several types of job titles, including: Bridge/Structure Inspection Team Leader, City Engineer, Civil Engineer, Civil Engineering Manager, County Engineer, Design Engineer, Project Engineer, Railroad Design Consultant, Structural Engineer, Traffic Engineer.

They would perform such tasks as:

Figure 49 Civil Engineers Inspecting a Dam (Source: USACE)

---

[18] National Center for O*NET Development. Civil Engineers. *17-2051.00*. Retrieved November 17, 2018, from https://www.onetonline.org/link/summary/17-2051.00

- Compute load and grade requirements, water flow rates, or material stress factors to determine design specifications.
- Test soils or materials to determine the adequacy and strength of foundations, concrete, asphalt, or steel.
- Manage and direct the construction, operations, or maintenance activities at project sites.
- Plan and design transportation or hydraulic systems or structures using computer assisted design or drawing tools.
- Prepare or present public reports on topics such as bid proposals, deeds, environmental impact statements, or property and right-of-way descriptions.
- Design energy efficient or environmentally sound civil structures.
- Direct engineering activities ensuring compliance with environmental, safety, or other governmental regulations.
- Analyze survey reports, maps, drawings, blueprints, aerial photography, or other topographical or geologic data.
- Conduct studies of traffic patterns or environmental conditions to identify engineering problems and assess potential project impact.
- Design or engineer systems to efficiently dispose of chemical, biological, or other toxic waste.

Mechanical Engineers

Mechanical devices transform energy in one form into another, such as fuel into heat, rotation into translation, steam into turbine drives, gasoline and air into automotive motion, etc. Mechanical Engineers perform engineering in planning and designing tools, engines, machines, and other mechanically functioning equipment. They also oversee installation, operation, maintenance, and repair of equipment such as centralized heat, gas, water, and steam systems.[19]

Mechanical Engineers often work under job titles such as: Application Engineer, Design Engineer, Design Maintenance Engineer, Equipment Engineer, Mechanical Design

Figure 50 Cassie, a walking robot built at Oregon State University Dept of Mechanical Engineering.

---

[19] National Center for O*NET Development. Mechanical Engineers. *17-2141.00*. Retrieved November 17, 2018, from https://www.onetonline.org/link/summary/17-2141.00

Engineer, Mechanical Engineer, Process Engineer, Product Engineer, Project Engineer, Test Engineer

They would perform such tasks as:

- Research, design, evaluate, install, operate, or maintain mechanical products, equipment, systems or processes to meet requirements.
- Develop, coordinate, or monitor all aspects of production, including selection of manufacturing methods, fabrication, or operation of product designs.
- Recommend design modifications to eliminate machine or system malfunctions.
- Design test control apparatus or equipment or develop procedures for testing products.
- Direct the installation, operation, maintenance, or repair of renewable energy equipment, such as heating, ventilating, and air conditioning (HVAC) or water systems.
- Design integrated mechanical or alternative systems, such as mechanical cooling systems with natural ventilation systems, to improve energy efficiency.
- Calculate energy losses for buildings, using equipment such as computers, combustion analyzers, or pressure gauges.
- Study industrial processes to maximize the efficiency of equipment applications, including equipment placement.
- Establish or coordinate the maintenance or safety procedures, service schedule, or supply of materials required to maintain machines or equipment in the prescribed condition.
- Select or install combined heat units, power units, cogeneration equipment, or trigeneration equipment that reduces energy use or pollution.

Electrical Engineers

Electrical systems transform electrical potential and power into motors, lights, telecommunications, controls and similar applications. Electrical Engineers perform research, design, develop, test, or supervise the manufacturing and installation of electrical equipment, components, or systems for commercial, industrial, military, or scientific use.

Electrical Engineers often work under job titles such as: Circuits Engineer, Design Engineer, Electrical Controls Engineer, Electrical Design Engineer, Electrical Engineer, Electrical Project Engineer, Instrumentation and Electrical Reliability Engineer (I&E Reliability Engineer), Power Systems Engineer, Project Engineer, Test Engineer.[20]

Figure 51 Wynton Habersham, chief electrical officer for MTA New York City Transit's subway system, showing control panel. (Photo: Metropolitan Transportation Authority / Patrick Cashin.)

They would perform such tasks as:

- Prepare technical drawings, specifications of electrical systems, or topographical maps to ensure that installation and operations conform to standards and customer requirements.
- Design, implement, maintain, or improve electrical instruments, equipment, facilities, components, products, or systems for commercial, industrial, or domestic purposes.
- Perform detailed calculations to compute and establish manufacturing, construction, or installation standards or specifications.
- Prepare specifications for purchases of materials or equipment.
- Plan or implement research methodology or procedures to apply principles of electrical theory to engineering projects.
- Design electrical systems or components that minimize electric energy requirements, such as lighting systems designed to account for natural lighting.
- Plan layout of electric power generating plants or distribution lines or stations.

---

[20] National Center for O*NET Development. Electrical Engineers. *17-2071.00*. Retrieved November 17, 2018, from https://www.onetonline.org/link/summary/17-2071.00

Chemical Engineers

Chemical plants refine and transform molecules in thousands of useful forms. Chemical Engineers design chemical plant equipment and devise processes for manufacturing chemicals and products, such as gasoline, synthetic rubber, plastics, detergents, cement, paper, and pulp, by applying principles and technology of chemistry, physics, and engineering.

Chemical Engineers often work under job titles such as: Chemical Engineer, Development Engineer, Engineer, Engineering Scientist, Process Control Engineer, Process Engineer, Project Engineer, Refinery Process Engineer, Research Chemical Engineer, Scientist

They would perform such tasks as[21]:

- Develop safety procedures to be employed by workers operating equipment or working in close proximity to ongoing chemical reactions.
- Develop processes to separate components of liquids or gases or generate electrical currents, using controlled chemical processes.
- Evaluate chemical equipment and processes to identify ways to optimize performance or to ensure compliance with safety and environmental regulations.
- Perform laboratory studies of steps in the manufacture of new products and test proposed processes in small-scale operation, such as a pilot plant.
- Prepare estimate of production costs and production progress reports for management.
- Design measurement and control systems for chemical plants based on data collected in laboratory experiments and in pilot plant operations.
- Determine most effective arrangement of operations such as mixing, crushing, heat transfer, distillation, and drying.
- Direct activities of workers who operate or are engaged in constructing and improving absorption, evaporation, or electromagnetic equipment.

Figure 52 A Chemical engineer stands in front of the boiler inside the Explosive Destruction System (EDS) Boiler Chiller Container at the Pueblo Chemical Agent-Destruction Pilot Plant. The boiler will provide steam during destruction operations. (Source: PEO ACWA)

---

[21] **National Center for O*NET Development.** 17-2041.00 - Chemical Engineers, accessed December 16, 2018, https://www.onetonline.org/link/summary/17-2041.00

- Perform tests and monitor performance of processes throughout stages of production to determine degree of control over variables such as temperature, density, specific gravity, and pressure.
- Design and plan layout of equipment.

Marine Engineers

Marine devices must float upon, travel across or withstand the sea. Marine Engineers design, develop, and take responsibility for the installation of ship machinery and related equipment including propulsion machines and power supply systems.

Marine Engineers often work under job titles such as: Consulting Marine Engineer, Hull Outfit Supervisor, Marine Consultant, Marine Design Engineer, Marine Engineer, Marine Engineering Consultant, Marine Surveyor, Project Engineer, Propulsion Machinery Service Engineer, Ships Equipment Engineer[22]

They would perform such tasks as:

- Perform monitoring activities to ensure that ships comply with international regulations and standards for life-saving equipment and pollution preventatives.
- Check, test, and maintain automatic controls and alarm systems.
- Evaluate operation of marine equipment during acceptance testing and shakedown cruises.
- Conduct environmental, operational, or performance tests on marine machinery and equipment.
- Inspect marine equipment and machinery to draw up work requests and job specifications.
- Design and oversee testing, installation, and repair of marine apparatus and equipment.
- Procure materials needed to repair marine equipment and machinery.

Figure 53 Crew members aboard the Auxiliary General Oceanographic Research (AGOR) vessel R/V Sally Ride retrieve scientific moorings located in La Jolla canyon. (U.S. Navy photo by John F. Williams/Released)

---

[22] **National Center for O*NET Development. Marine Engineers.** *17-2121.01.* Retrieved November 17, 2018, from https://www.onetonline.org/link/summary/17-2121.01

- Maintain and coordinate repair of marine machinery and equipment for installation on vessels.
- Conduct analytical, environmental, operational, or performance studies to develop designs for products, such as marine engines, equipment, and structures.
- Review work requests and compare them with previous work completed on ships to ensure that costs are economically sound.

Summary of Engineering

Table 15 Compares selected engineering disciplines in terms of their major areas of designed product or services and the scientific principles applied or used by their technologies. This comparison can be expanded to including the hundreds of existing and evolving

Table 15 Comparison of Selected Engineering Disciplines

Engineering Discipline	Typical Products and Services Designed	Major Principles Applied
Civil	Structures and Facilities on Earthen Foundations for Public Needs	Soil Behavior, Material Strength, Hydraulics, Structural Strength, etc.
Mechanical	Energy Transformation and motion through mechanisms and engines.	Kinematics, Dynamics, Vibrations, Thermodynamics, Machine Element Analysis, etc.
Chemical	Re-Arrangement of Molecules on an Industrial Scale, with storage and packaging.	Physics, Chemistry, Hydraulics, Thermodynamics, Corrosion, Safety, Unit Processes, etc.
Electrical	Energy Transformation to/from Electricity, Magnetism and Light for public and commercial applications	Classical and Modern Physics, Power Generation/Transmission, Antenna, Semiconductors, Controls, Sensors, etc.
Marine	Vessels and Structures that float on or are heavily affected by large bodies of water	Lakebed and Seabed Soils, Structures, Sea States and Wave Loading, Dynamics, Vessel Stability, etc.

disciplines and sub-disciplines that specialize in applying our ever-changing scientific knowledge and areas of need.

One important take-away from this comparison is that for every new set of requirements, there is usually a principle or technology that may apply. Engineers are always out looking for new ideas, products and services that might solve a problem. It is a dynamic profession.

Design Technician disciplines

Designers is the current name for "drafters", an obsolete term. Designers apply techniques, methods known to produce the desired result when implemented. Frequently responsible for the "first draft" of the design, the proposed layout or design is then checked by several engineering disciplines.

Technician refers to specializing in the use of existing technologies, i.e., products and services for which the performance is known. As technicians, the designers' analysis is standardized, which allows them to consider more functions simultaneously. Designers check for:

- Standardization of Components (vs Specialty)
- Constructability/Manufacturability/Fabricability
- Accessibility
- Operability
- Maintainability
- Assembly

These are not sciences, per se, but they are principles for which the design must be checked. Many of these principles are learned on the job and not in formal classes.

Vendors/Suppliers

The great variety of vendor-supplied products often makes purchasing a component much cheaper than designing and building it. When standards exist for components, equipment and materials, vendors prefer to offer products that meet or exceed the performance required by the standard.

In short, vender supplied products are usually cheaper and come with an analysis of performance. They also, however, imposed new requirements of their own.

Factory/Shop/Construction disciplines and trades

During the design itself, input is usually sought from the people who have built similar designs. They can quickly and easily assure that a design can be built economically and using the tools and equipment available. The savings is usually realized during the implementation part of the work package. In other words, design should always "anticipate" the implementation even if the designers have not be charged with designing it.

Manufacturing Occupations

Manufacturers are mainly concerned with products that are self-contained and function as completed at the factory.

Manufacturing Occupations include:

- Computer numerically controlled machine tool programmers and operators

Figure 54 Sandblasting before painting (source: US Air Force)

- Machine setters, operators and tenders
- Computer-controlled machine tool
- Painters
- Machinists
- Tool and die makers
- Engine and other machine assemblers
- Industrial machinery mechanics
- Welders, cutters, solderers and brazers
- Millwrights
- Fiberglass laminators and fabricators

Fabrication Shop Trades

Fabricated parts generally require further assembly elsewhere to function as intended. In most heavy civil or plant construction, fabrication of major parts in a shop can save a lot of "on-site" labor. Much equipment, vessels, major components, sub-systems, etc., are fabricated and tested in a shop before being shipped to the job site for final installation. Work package designers will try to consider what tools and skills are available in such shops to reduce cost and schedule.

Figure 55 Worker uses cutting torch. (Source: US Air Force)

Typical Fabrication Shop occupations include:

- Steel erector, also known as an iron turtle
- Welder
- Boilermaker
- Pipefitter
- Millwright
- Laborer

Light Construction Trades

Light construction is dominated by trades that bring their own tools. Most labor is expended on-site although the trend is to pre-fabricate major sub-assemblies elsewhere and assemble on-site.

Light Construction trades include such jobs as:

- Carpenter
- Carpet Installer
- Cement & Concrete Finisher

- Fencer/Fence Erector
- Flooring Installer
- Glazier
- HVAC Tech
- Insulation Worker
- Landscaper
- Mason
- Painter
- Plasterer
- Plumber
- Roofer

Figure 56 Stardust Industries light construction worker (USDA Photo by Lance Cheung)

Heavy Construction Trades

Heavy Construction involve larger quantities of materials, more hours of labor, higher use of heavy machinery. Although specialty trades bring most of their own tools, the constructor supplies the heavy equipment and arranges for equipment and sub-assemblies to be built off-site.

Heavy Construction trades include such jobs as:

- Boilermaker
- Dredge Operator
- Electrician/Technician
- Equipment Operator
- Elevator Mechanic
- Estimator
- Foreman
- Iron-worker
- Laborer
- Millwright
- Pile Driver Operator
- Pipefitter, Steamfitter
- Sheet Metal Worker
- Safety Manager
- Construction Manager
- Ironworker
- Safety Manager
- Steel erector, also known as an iron turtle
- Welder

Figure 57 Setback levee construction in West Sacramento, CA. (U.S. Army Corps of Engineers photo by Michael J. Nevins)

- Rigger
- Heavy equipment operator
- Shipping-and-receiving workers

Outsourcing Implementation

Outsoucing is the delegation of parts of the design and implementation to others, through purchases, sub-contracts and joint ventures. Often, creation of the implementation portion of a work package is generally left to manufacturers, fabricators, and constructors. Even then, they a frequently asked to check the designs for correct used of available skills, equipment and facilities.

Most of the non-critical design details are left to shop/construction disciples and trades. Overall requirements, in the form of schematics, general arrangements, connection lists, etc. are delegated out to these professionals for implementation. They are free to design and implement to physical routings and connections as they see fit, as long as the overall requirements are met.

However, a purely made-to-order work package involves no design at all. There are no functional requirements present and no new products designed and proven to meet them. The trades involved mere perform the work prescribed to the quality standards set in the work package.

Operations & Maintenance Disciplines

Operations & Maintenance (O&M) refers to the post-implementation phase, when the design operates safely within normal parameters and earns the returns anticipated. Designers seek input is sought from those who have operated and maintained similar designs. They have had to live with prior design mistakes and can pass those lessons on to designers and their designs. These lessons, usually in the form of O&M requirements on designs, improve operability and maintainability.

Figure 58 Operations and Maintenance (O&M) personnel often advise designers on best practices. (Photo: US ACWA)

Many of the requirements of the end users and maintainers are "unwritten" requirements. Ugly, hard-to-use, hard-to-maintain things don't sell as well. Just because the requirements aren't written, doesn't mean they are not there. The best designers come from manufactures, fabricators, assemblers, operators, maintainers and others familiar with features that help or hurt their work.

Research, Development and Testing Disciplines

Finally, the scientific principles, techniques, inventions, etc., are developed by organizations specializing in that research. These types of research include:

- **Basic Research.** Uncovering principles without regard to their eventual application.
- **Applied Research.** Searching for principles with eventual applications as the goal.
- **Development.** Identifying and demonstrating that specific principles apply to a product that can be implemented. Prototypes and pilot implementations are examples of development.
- **Testing** is a confirming step that a product performs as expected.

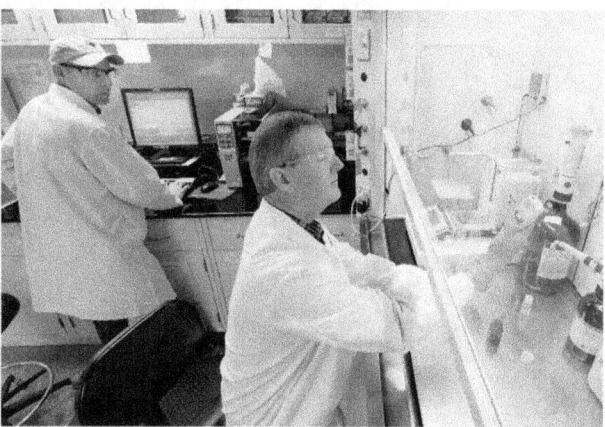

Figure 59 Research develops technologies that are applied in design. (Photo: US ACWA)

These are the sources of technologies applied in designs everywhere. Their work is essential to progress in our world.

Chapter Take-Aways

All STEM areas contribute to design in various ways as do other, non-scientific disciplines. Engineers scientific principles and technologies that use them. Scientists discover principles to apply and Mathematicians proved the tools needed to express those principles in useful form.

Other disciplines contribution to design. Vendors/Suppliers provide products with rated performance and additional analysis to prove they meet design requirements. Implementation disciplines verify the designs can be manufactured, fabricated and constructed. Maintenance and operations can predict usability and upkeep in the field.

Finally, Research, Development and Testing disciplines discover scientific principles and develop them into performance-rated products and services that can be used in designs.

Problems

1. Of the four types of STEM fields (Science, Technology, Engineering and Mathematics), which one develops principles or tools used by other fields? Which fields apply these tools and principles?
2. If you like to discover how the world works, which STEM fields should you consider?

3. If you like to "re-arrange the formula" to get a simple answer, which STEM fields should you consider?
4. If you like to build technical things, which STEM fields should you consider?
5. Why are some products needing a lot of technical advice?
6. Why are plumbers, electricians and carpenters hired to advise customers in hardware stores?
7. Why would drug store customers consult the pharmacists before taking prescription and non-prescription products?

Conclusions

We have explained design in simple terms. Design is really a part of our everyday lives. This book adds a formal structure to what you already do.

Design is so simple that we explained it in the first few pages. Later, we added more formality to design itself by discussing specifications and the design process itself.

There is nothing in this book that restricts design to engineers, only. Many fine designers apply principles other than physics and chemistry in their domain of practice. They are specialists in their own fields. They are not always called designers but design is part of what they do.

Sources Used

I've used many images and art from public domain or GNU licensed sources. These sources are noted in footnotes and captions. The main sources are:

Wikipedia https://www.wikipedia.org/

Wikimedia Commons https://commons.wikimedia.org/wiki/Main_Page

Publicdomainvectors.org https://publicdomainvectors.org/

Flickr (public domain area) https://www.flickr.com/

Glossary

Term	Definition
Abstraction	A view of design leaving out irrelevant details.
Acceptance Tests	A battery of trials confirming expected functional performance of a design.
Analysis	The process of breaking a complex topic or substance into smaller parts in order to gain a better understanding of it
Applied Research	Searching for principles with eventual applications as the goal.
Applied Science	Taking known scientific principles discovered in the laboratory and scale them up to to predict design performance.
Approval	Consent to a formal requests that a choice or change be accepted as baseline
As-Built	The configuration of the system after it is built, accepted and operational.
Baseline	Design characteristics that have been approved for use in all subsequent efforts.
Basic Research	Studies to uncover principles without regard to their eventual application.
Bill of Materials	Physical inputs stating sizes, quantities and characteristics of items required.
BOM	see Bill of Materials
Business Case	An estimated return on the product or service that satisfies the market need
CAD	Computer-Aided-Design
CAPEX	See Capital Expenditures
Capital Expenditures	Expenses that create a capital assets and place it into production. CAPEX costs are depreciated or amortized over time.
Chemical Engineers	The design chemical plant equipment and processes for manufacturing chemicals and products, such as gasoline, synthetic rubber, plastics, detergents, cement, paper, and pulp.
Civil engineering	The design, construction, and maintenance of the physical and naturally built environment, including works such as roads, bridges, canals, dams, airports, water and wastewater systems, pipelines, and railways.
Computer-Aided-Design	A computer system concentrating on capturing and storing the data as needed by the application
Context, of Design	A series of analyses linking a Market Need through Functional Requirements.
Controls (of a Work Package)	Tools used to assure proper execution of the work package processes.
Convergence	The narrowing of undecided options toward a completed defined acceptable design.
COTS	Commercial-Off-The-Shelf
Criterion	The standard value against which the design characteristic is judged, and
Data of Record	The location where the authoritative data value is stored.
Decomposition	A view of a system as its subsystems, that together form the complete system.

Descriptive Geometry	The representation of three-dimensional objects in two dimensions by using a specific set of procedures
Design	Definition of a product or service with proof that it peforms as specified in a Functional Requirement.
Design Basis	At the beging of design the current approved Requirements, Baseline and any additional constraints on the work.
Design Procedures	Steps leading to designs with proven performance
Design Technology	The design and assurance of non-scientific requirements, including, Standardization of Components (vs Specialty), Constructability/Manufacturability/Fabricatability, Accessibility. Operability, Maintainability and Assembly,
Designers	Specialists skilled at defining a solution and proving their required functionality. Frequently, they also design the work package that will create or implement the defined solution.
Development	Identifying and demonstrating that specific principles apply to a product that can be implemented. Prototypes and pilot implementations are examples of development.
Dimensional Analysis	The reduction the number of variables to study in a phenomenon to a minimal number of dimensionless numbers to minimize lab testing.
Electrical Engineering	The research, design, develop, test, or supervision of the manufacturing and installation of electrical equipment, components, or systems.
End Product (of a Work Package)	The main output of a work package designed to create it.
Engineers	Specialists in appling known scientific principles to their design to predict technical performance.
Essential Implementation Procedures	Implemtation procedures that were required in the analyses proving the design.
Essential Parameter	A parameter used in proving a design meets its functional requirements or is required for implementation.
Feasibility	The question of whether the requirements can be satisfied before consuming allocated resources.
Functional Requirement	What the design object must do.
Functional Requirement (of a product or process)	A product or process defined by the functions it must perform
Gate	A go/nogo decision that can halt the processes or allow it to continue.
IDEF0	A comprehensive method of documenting processes. IDEF0, a compound acronym ("Icam DEFinition for Function Modeling", where ICAM is an acronym for "Integrated Computer Aided Manufacturing"),
Implementation (of a design)	The work package that creates the defined product

Term	Definition
Impossibility	The question of whether the requirements can be satisfied without violating a physical or regulatory law.
Inputs (of a Work Package)	The items that are "processed" into the work package outputs.
Inspection	Measurement of the characteristic on a product and comparison it to the criterion specified.
Interval Scale	Interval scales have all the characteristics of ordinal scales but have a standard interval separating each item in the sequence.
Inversion	The use of the inverse of a performance function to calculate a design characteristic from required peformance.
Iteration	Repetitive change and re-assessment of system perform until all the requirements are satisfied.
Management Plan	Explaination of how the designer will show progress and deliver final results
Marine Engineering	The design, development and installation of ship machinery and related equipment including propulsion machines and power supply systems.
Market Need	A generalized statement, with supporting data, that a need exists for a product or service strong enough for it to be sold
Measure	The parameter that is to be a characteristic of the design,
Mechanical Engineers	The planning and design of tools, engines, machines, and other mechanically functioning equipment.
Nominal Scale	Assigning the measured quantity to a name with noparticular order.
O&M	See Operations & Maintenance
Operational Expenditures	Operations and Maintenance expenses of an existing capital assets after production begins. OPEX is subtracted from gross income calculate gross profit.
Operations & Maintenance	The post-implementation phase, when the design operates safely within normal parameters and earns the returns anticipated.
Operator	The comparison to be made between the measure and criterion.
OPEX	Operational Expenditures
Ordinal Scale	Ordinal scales have all the attributes of nominal scales but add order.
Outputs (of a Work Package)	The defined results of a completed work package
Outsoucing	The delegation of parts of the design and implementation to others, through purchases, sub-contracts and joint ventures.
Phasing	Breaking up the design work into logical periods to allow an estimate of remaining requirements on resources and to re-evaluate the business case.
Pilot	A smaller (pilot) version of the end system built and subjected to extensive testing.
Pre-Design Processes	Studies and steps designed to reduce risk and provide oversight of actual design work
Prescriptive Product Definition	An item defined by its measurable characteristics

Term	Definition
Procedure	For activities, specifies for each step what needs to be done, when, and by whom. (See Work Package)
Process Requirements	See Processes (of a Work Package)
Processes (of a Work Package)	The steps to be taken to transform the inputs into outputs.
Product Definition	see Prescriptive Product Definition
Product Specs	See Rated Performance
Propagation of constraint	The incorporation of design choice consequences into all other parts of the design.
Prototype	A preliminary (prototype) version of the end system is built and subjected to extensive testing.
Rated Peformance	Publish minimal values of stated measures from suppliers that can be relied on for comparison to required specifications.
Ratio Scales	The Ratio Scale is an interval scale with a true zero.
Recipe	The simplest form of work package used to prepare foods
Refinement	The addition of detail to a design, or in other words, making a design more concrete.
Rendering	The generation of a view or report completely from existing data.
Representation(s)	Displays of data for entry and review.
Requirements Engineering	The process of defining, documenting and maintaining requirements in the engineering design process.
Resource Requirements	See Resources (of a Work Package)
Resources (of a Work Package)	The products and services that are used by the work package processes.
Resources Budget	The resources, e.g., money and schedule time, the designer can use to complete the design
Rollup	Determining the impact on high-level requirements of low-level design changes.
Scales	Scale of measure is a classification that describes the nature of information within the values assigned.
Scope of Work	The level of detail of design to be accomplished and of the implementation work package, if included
Similitude	The techique of using carefully scaled models to predict performance of real systems.
Solution	The defined and proven product or service that satisfies the functional requirements
SOP	See Standard Operating Procedure
specification clause	A measure, a criterion and an operator that logically true or false (satisfied or not-satisfied). The criterion is in the same scale and the operator must be appropriate to that scale.

Standard Operating Procedure	Pre-defined prodcedures to help workers carry out complex routine operations.
Standard Unit of Measure	An agreed unit that has a physical standard or repeatable defining derivation that everyone can replicate for their own use
STEM	Science, Technology, Engineering and Math
Systems Modeling	Predicting the performance of systems made up of many individual components interacting in complex ways.
Technology	A product or process for which the performance is known
Testing	Confirming step that a product performs as expected.
Tolerance	The range of allowed results establishes the on the measurement.
Trades	The localized design and crafting of implementations in factories, fabrication shops and construction sites.
Venn Diagram	A diagram of all possible thing in the Universe (U) and the sets of things we define. In design, we are interested in the set of acceptable solutions.
Work Package (definition of item)	An item defined by a process that creates it
Work Package Elements:	A work package has 5 elements: 1. Inputs Required, 2. Process(s), 3. Resources Required, 4. Controls, and 5. Output.

Index

Acceptable Set, 31
Acceptance Tests, 17
Analysis, 53
Approvals, 57
As-Built, 58
Bill of Materials, 12
BOM. *See* Bill of Materials
Business Case, 16
CAD. *See* Computer-Aided-Design
Checklists, 56
Commercial Off The Shelf. *See* COTS
Computer-Aided Drafting, 50
Computer-Aided-Design, 50
Confirmed by Inspection, 11
Constraints on Implementation, 54
COTS, 18, 55
Criteria, 28
Criterion, 8, 28
Data of Record, 51
Descriptive Geometry, 50
Design Basis, 49
Design Context, 16
Design Procedures, 56
Designers, 18
Digital Twin, 58
Drawing, 50
Elements of a Work Package.
Essential Parameter, 50
Functional Requirement, 18
Functional Requirements, 5, 15, 16
GO/NOGO gauge, 29
Interval Scale, 28
Level of Measurement. *See* Scale of Measure
Management Plan, 49
Market Need, 16
Measure Advanced Operations, 30
Measure Operators, 29
Measure Property, 29

Measurement Scale Comparison, 29
Measure's Central Tendency, 30
Nominal Scale, 27
Nominal Scales, 24
Nominal Sizes, 9
Ordinal Scale, 9, 27
Ordinal Scales, 25
Pilot Test. *See* Prototype Test
Prescriptive Product Definition, 5, 9
Procedure. *See* Work Package
Process Definition, 32
Product Definition, 16, 18, **See** Prescriptive Product Definition
Propagation of Constraint, 54
Prototype Test. *See* Pilot Test
Ratio Scale, 27, 28
Rendering, 52
Rollup, 54
Scale of Measure, 24
Scope of Work, 49
Solution, 35
Specification, 30, 31,
Specification Clause, 23,
Specifications.
Standard Unit of Measure, 7
Systems Approach to Design, 17
Technology, 36, 55
Tolerance, 8, 29
Unfeasible, 37
Universal Set, 31
Venn Diagram, 31
Work Package, 5, 11, 16, 32, 35
Work Package Controls, 13
Work Package Inputs.
Work Package Outputs, 14
Work Package Processes.
Work Package Resources, 13

Appendix A – Writing a Design Problem Statement

Using the ideas of this book, you can write a technical problem statement. For the purposes of this discussion, we'll leave out all the commercial details. This problem statement is written by the client without regard to any particular design solution or designer.

Here is a short outline of what should be included:

- **Background** – Information necessary to understanding the need.
- **Functional Requirements** – The current Functional Requirements as known today as lists of logical clauses.
- **Design Basis** – The approved state of the design including the supporting analysis by the client and others.
- **Possible Solutions** – These list solutions that the client wants considered and those it has already rejected (and why). Any restrictions on technology use should also be noted.
- **Scope of Work** – The product definitions and analyses expected during the project and the level of detail expected at a successful conclusion.
- **Resources Available** – Money, schedule, facilities, expertise, software and other resources that are made available by the client on an as-needed, as-approved basis.
- **Reporting Requirements** – The interim/final reports and documents expected, usually referred to as "deliverables." These normally give the formats expected and the mode of delivery, e.g., "uploaded to the project server."

Appendix B – Writing a Technical Proposal

Once supplied with a problem statement, the designer usually replies with a proposal. Leaving out all the commercial details the technical proposal should contain:

- **An Introduction to the Designer** – A summary of why the designer is qualified to define products meeting this type of need and their experience with analyzing their performance.
- **Proposed Modifications to the Functional Requirements**. The changes and exceptions needed by the designer to feasibly perform the work. A rationale for each change should be given.
- **Approach to Extending the Current Baseline** – How the designer plans to subdivide and add detail to the systems and subsystems of the current baseline. This usually includes a Work Breakdown Structure (WBS) and a technical management plan for handling design decision consequences.
- **Designer Resources Available to the Work** – These include key personnel, facilities, software and hardware.
- **Analysis of Design Risks and Their Mitigation** – How the design will minimize the feasibility risk by careful management of resources.

Appendix C – Writing a Final Technical Report

When the design finishes successfully, a final report is submitted. Here is a short outline of its contents:

- **Introduction** – A short history of the project and summary of major accomplishments
- **Final Functional Requirements** – The Final Functional Requirements as approved by the client. Major modifications from the initial set should be noted and briefly explained.
- **Final Design Baseline** – The Final Approved Baseline and, if implemented, with full as-built updates. Frequently, the documents are so numerous that they submitted separately and summarized here.
- **Analysis of Design Performance** – The analysis of the design performance with rollup to the functional requirements. If acceptance testing confirmed the analysis, their results should be summarized as well. Show the satisfaction of every Functional Requirement.
- **Implementation Requirements** - Essential Implementation Parameters and Procedures should be listed to as to be including in the implementation phase.
- **Key Value Delivered** - Lessons learned beyond satisfying the requirements should be stated. Any extra performance, project savings, intellectual property or other value delivered to the client should be summarized here.
- **Opportunities for Future Work** - Any possible extensions that should be considered should be stated.

About the Author

William G. Beazley, Ph.D., PE

Dr. Beazley has been teaching and practicing design for over 40 years. In diverse fields from Mechanical Engineering, General Engineering through Piping Design Dr. Beazley has used a unique blend of learning theory and STEM disciplines to make design understandable. Thousands have taken his courses or viewed his training videos.

Dr. Beazley received his MS and PhD degrees in Mechanical Engineering from the University of Texas at Austin. He received his BS degree from the Tulsa University with dual majors in Mechanical Engineering and Psychology.

Dr. Beazley has worked on computer/web-based work practices, task analysis, automation and integration in the Petrochemical and Aerospace industries. He has taught and developed curriculum at universities, community colleges, professional societies and training centers. He contributed to several national standards, including IGES, STEP, EDI, and XML. He has developed content in Angel, Blackboard, Moodle using PowerPoint, video, Open Badges.

Dr. Beazley taught Mechanical Engineering Senior Design at the University of Texas and Freshman Design at the University of Illinois. At both, he developed criterion-based curriculum and assessments.

Dr. Beazley is best known for developing training and testing for Professional Piping Designers as Executive Director of the Society of Piping Engineers and Designers (SPED). SPED made history in describing successive levels of piping designer skills from pipe route to senior lead. Thousands have taken his training and PPD Certification exams.

NOTES

NOTES

www.ingramcontent.com/pod-product-compliance
Lightning Source LLC
Chambersburg PA
CBHW081604220526
45468CB00010B/2760